PERGAMON INTERNATIONA
of Science, Technology, Engineering an
The 1000-volume original paperback library
industrial training and the enjoyment of leisure
Publisher: Robert Maxwell, M.C.

INTRODUCTION TO SUPERCONDUCTIVITY
SECOND EDITION

THE PERGAMON TEXTBOOK
INSPECTION COPY SERVICE

An inspection copy of any book published in the Pergamon International Library will gladly be sent to academic staff without obligation for their consideration for course adoption or recommendation. Copies may be retained for a period of 60 days from receipt and returned if not suitable. When a particular title is adopted or recommended for adoption for class use and the recommendation results in a sale of 12 or more copies, the inspection copy may be retained with our compliments. The Publishers will be pleased to receive suggestions for revised editions and new titles to be published in this important International Library.

INTERNATIONAL SERIES IN SOLID STATE PHYSICS
Volume 6
GENERAL EDITORS: R. SMOLUCHOWSKI and N. KURTI

OTHER TITLES OF INTEREST

BASSANI & PASTORI PARRAVICINI:
Electronic States and Optical Transitions in Solids

BUSCH & SCHADE:
Lectures on Solid State Physics

HAUG:
Theoretical Solid State Physics (2 vols)

HOLT & HANEMAN:
Defects and Surfaces in Semiconducting Materials and Devices

JARZEBSKI:
Oxide Semiconductors

ROY:
Tunnelling and Negative Resistance Phenomena in Semiconductors

SHAY & WERNICK:
Ternary Chalcopyrite Semiconductors: Growth, Electronic Properties and Applications

TANNER:
X-Ray Diffraction Topography

WILLIAMS & HALL:
Luminescence and the Light Emitting Diode

The terms of our inspection copy service apply to all the above books. Full details of all books listed and specimen copies of journals listed will gladly be sent upon request.

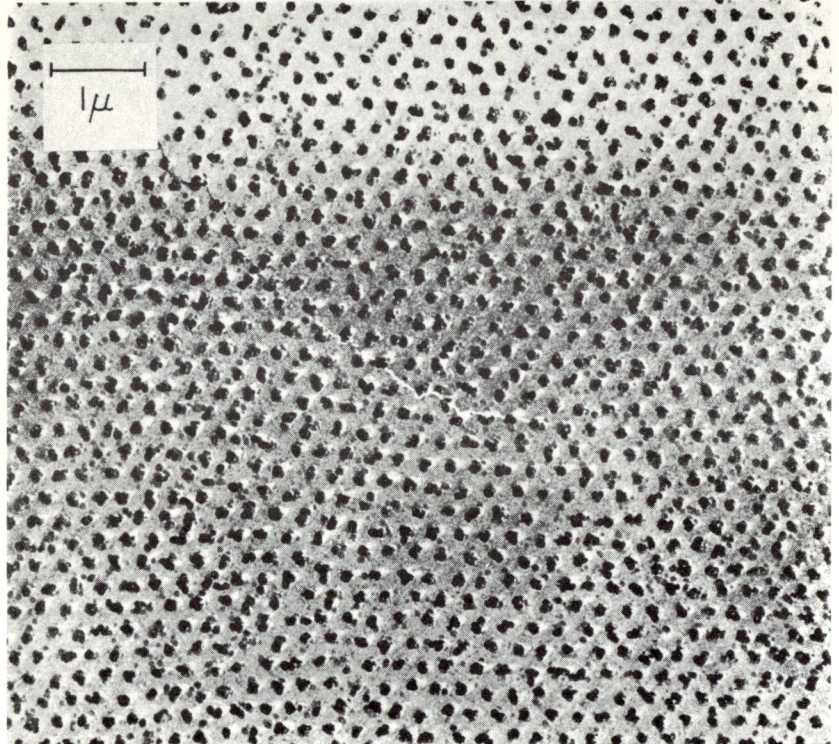

PATTERN OF INDIVIDUAL FLUXONS IN A TYPE-II SUPERCONDUCTOR

This photograph shows the triangular pattern of fluxons in a type-II superconductor (see Chapter 12). The pattern is revealed by allowing very small (500 Å) ferromagnetic particles to settle on the surface of a magnetized specimen (lead–indium alloy). The particles locate themselves where the magnetic flux intersects the surface. The photograph was obtained by electron microscopy of the deposited particles. (Photograph by courtesy of V. Essmann and H. Träuble, Max Plank Institut für Metallforschung.)

INTRODUCTION TO SUPERCONDUCTIVITY

BY

A. C. ROSE-INNES

Professor of Physics and Electrical Engineering

AND

E. H. RHODERICK

Professor of Solid-State Electronics
University of Manchester Institute of Science and Technology
U.K.

SECOND EDITION

PERGAMON PRESS

OXFORD · NEW YORK · TORONTO · SYDNEY · PARIS · FRANKFURT

U.K.	Pergamon Press Ltd., Headington Hill Hall, Oxford OX3 0BW, England
U.S.A.	Pergamon Press Inc., Maxwell House, Fairview Park, Elmsford, New York 10523, U.S.A.
CANADA	Pergamon of Canada Ltd., 75 The East Mall, Toronto, Ontario, Canada
AUSTRALIA	Pergamon Press (Aust.) Pty. Ltd., 19a Boundary Street, Rushcutters Bay, N.S.W. 2011, Australia
FRANCE	Pergamon Press SARL, 24 rue des Ecoles, 75240 Paris, Cedex 05, France
FEDERAL REPUBLIC OF GERMANY	Pergamon Press GmbH, 6242 Kronberg-Taunus, Pferdstrasse 1, Federal Republic of Germany

Copyright © 1978, Pergamon Press Ltd.

All Rights Reserved. No part of this publication may be reproduced, stored in a retrieval system or transmitted in any form or by any means: electronic, electrostatic, magnetic tape, mechanical, photocopying, recording or otherwise, without permission in writing from the publishers

First edition 1969

Reprinted with corrections 1976

Second edition 1978

Library of Congress Cataloging in Publication Data

Rose-Innes, Alistair Christopher.
Introduction to superconductivity.

(International series in solid state physics; v.6)
Includes index.
1. Superconductivity. I. Rhoderick, E. H., joint author. II. Title.
QC612.S8R6 1977   573.6'23   77-4811

ISBN 0-08-021651-X   Hard cover
ISBN 0-08-021652-8   Flexi cover

Printed in Great Britain at The Pitman Press, Bath

CONTENTS

Preface to the Second Edition	ix
Preface to the First Edition	xi
Symbols	xiii
Introduction	xvii

PART I. TYPE-I SUPERCONDUCTORS

Chapter 1. Zero Resistance	3
1.1 Superconducting Transition Temperature	5
1.2 Zero Resistance	8
1.3 The Resistanceless Circuit	8
1.4 A.C. Resistivity	12
Chapter 2. Perfect Diamagnetism	16
2.1 Magnetic Properties of a Perfect Conductor	16
2.2 Special Magnetic Behaviour of a Superconductor	19
2.2.1 Meissner effect	19
2.2.2 Permeability and susceptibility of a superconductor	21
2.3 Surface Currents	22
2.3.1 Hole through a superconductor	24
2.4 Penetration Depth	26
2.4.1 Variation with temperature	28
Chapter 3. Electrodynamics	31
3.1 Consequence of Zero Resistance	31
3.2 The London Theory	33
3.2.1 An application of the London theory	38
Chapter 4. The Critical Magnetic Field	40
4.1 Free Energy of a Superconductor	41
4.2 Variation of Critical Field with Temperature	43
4.3 Magnetization of Superconductors	46
4.3.1 "Non-ideal" specimens	47
4.4 Measurement of Magnetic Properties	48
4.4.1 Measurement of flux density	49
4.4.2 Measurement of magnetization	50
4.4.3 Integrating method	52

v

Chapter 5. Thermodynamics of the Transition — 54

5.1 Entropy of the Superconducting State — 54
5.2 Specific Heat and Latent Heat — 57
    5.2.1 First-order and second-order transitions — 57
    5.2.2 Adiabatic magnetization — 59
    5.2.3 Lattice and electronic specific heats — 59
5.3 Mechanical Effects — 61
5.4 Thermal Conductivity — 62
5.5 Thermoelectric Effects — 63

Chapter 6. The Intermediate State — 64

6.1 The Demagnetizing Factor — 64
6.2 Magnetic Transitions for $n \neq 0$ — 67
6.3 The Boundary Between a Superconducting and a Normal Region — 68
6.4 Magnetic Properties of the Intermediate State — 69
6.5 The Gibbs Free Energy in the Intermediate State — 70
6.6 The Experimental Observation of the Intermediate State — 72
6.7 The Absolute Size of the Domains: the Role of Surface Energy — 74
6.8 Restoration of Resistance to a Wire in a Transverse Magnetic Field — 75
6.9 The Concept of Coherence and the Origin of the Surface Energy — 77

Chapter 7. Transport Currents in Superconductors — 82

7.1 Critical Currents — 82
    7.1.1 Critical currents of wires — 83
7.2 Thermal Propagation — 86
7.3 Intermediate State Induced by a Current — 89

Chapter 8. The Superconducting Properties of Small Specimens — 92

8.1 The Effect of Penetration on the Critical Magnetic Field — 92
8.2 The Critical Field of a Parallel-sided Plate — 93
8.3 More Complicated Geometries — 97
8.4 Limitations of the London Theory — 98
8.5 The Ginzburg–Landau Theory — 101
8.6 Edge Effects — 104
8.7 Transitions in Perpendicular Magnetic Fields — 105
8.8 Critical Currents of Thin Specimens — 106
8.9 Measurements of Critical Currents — 110

Chapter 9. The Microscopic Theory of Superconductivity — 112

9.1 Summary of the Properties of the Superconducting State — 112
    9.1.1 Zero resistance — 112
    9.1.2 Crystal structure — 113
    9.1.3 Electronic specific heat — 114
    9.1.4 Long-range order — 114
    9.1.5 The isotope effect — 115
    9.1.6 The Meissner effect — 115
9.2 The Concept of an Energy Gap — 115
9.3 The Bardeen–Cooper–Schrieffer Theory — 117
    9.3.1 Restatement of the problem — 117
    9.3.2 The electron–lattice interaction — 117
    9.3.3 Cooper pairs — 120
    9.3.4 The superconducting ground state — 125

CONTENTS

9.3.5	Properties of the BCS ground state	128
9.3.6	Macroscopic properties of superconductors according to the BCS theory	131
9.3.7	The current-carrying states	135
9.3.8	The pair wavefunction: long-range coherence	138

CHAPTER 10. TUNNELLING AND THE ENERGY GAP — 140

10.1	The Tunnelling Process	140
10.2	The Energy Level Diagram for a Superconductor	142
10.3	Tunnelling Between a Normal Metal and a Superconductor	143
10.4	Tunnelling Between Two Identical Superconductors	145
10.5	The Semiconductor Representation	147
10.6	Other Types of Tunnelling	149
10.7	Practical Details	150

CHAPTER 11. COHERENCE OF THE ELECTRON-PAIR WAVE; QUANTUM INTERFERENCE — 153

11.1	Electron-pair Waves	153
	11.1.1 Phase of the electron-pair wave	154
	11.1.2 Effect of a magnetic field	155
11.2	The Fluxoid	156
	11.2.1 Fluxoid within a superconducting metal	159
11.3	Weak links	160
	11.3.1 Josephson tunnelling	160
	11.3.2 Pendulum analogue	162
	11.3.3 a.c. Josephson effect	165
	11.3.4 Coupling energy	167
	11.3.5 Weak-links	169
11.4	Superconducting Quantum Interference Device (SQUID)	170
	11.4.1 "Diffraction" effects	178

PART II. TYPE-II SUPERCONDUCTIVITY

CHAPTER 12. THE MIXED STATE — 183

12.1	Negative Surface Energy	185
12.2	The Mixed State	186
	12.2.1 Details of the mixed state	188
12.3	Ginzburg–Landau Constant of Metals and Alloys	190
12.4	Lower and Upper Critical Fields	191
	12.4.1 Lower critical field, H_{c1}	191
	12.4.2 Upper critical field, H_{c2}	192
	12.4.3 Thermodynamic critical field, H_c	193
	12.4.4 Value of the upper critical field	194
	12.4.5 Paramagnetic limit	195
12.5	Magnetization of Type-II Superconductors	197
	12.5.1 Determination of $\varkappa$	198
	12.5.2 Irreversible magnetization	199
12.6	Specific Heat of Type-II Superconductors	200

CHAPTER 13. CRITICAL CURRENTS OF TYPE-II SUPERCONDUCTORS — 202

13.1	Critical Currents	202
13.2	Flow Resistance	204

13.3	Flux Flow		206
	13.3.1 Lorentz force and critical current		206
	13.3.2 Flux flow		212
	13.3.3 E.M.F. due to core motion		214
13.4	Surface Superconductivity		217

APPENDIX A. The Significance of the Magnetic Flux Density B and the Magnetic Field Strength H — 221

    A.1 Definition of B — 221
    A.2 The Effect of Magnetic Material — 222
    A.3 The Magnetic Field Strength — 224
    A.4 The Case of a Superconductor — 225
    A.5 Demagnetizing Effects — 227

APPENDIX B. Free Energy of a Magnetic Body — 230

INDEX — 233

PREFACE TO THE SECOND EDITION

THIS edition differs from the first edition chiefly in Chapter 11, which has been almost completely rewritten to give a more physically-based picture of the effects arising from the long-range coherence of the electron-waves in superconductors and the operation of quantum interference devices. We are very grateful to Dr. J. Lowell for reading and commenting on the draft of this rewritten chapter. There are a number of other relatively minor changes throughout the book which we hope will improve the presentation.

We must also thank Dr. Lucjan Sniadower for pointing out a number of errors, Miss Dorothy Denton for typing the manuscript of this second edition, and Nicholas Rose-Innes for help in checking the manuscript and proofs.

University of Manchester E.H.R.
  Institute of Science and Technology A.C.R.-I.
  December 1976

PREFACE TO THE FIRST EDITION

THIS book is based to a large extent on lectures we have given to undergraduate and first-year postgraduate students. We intend that it should indeed be an *introduction* to superconductivity and have selected and presented our material with this in mind. We do not intend the contents to be read as a definitive and exhaustive treatment of superconductivity; several such texts are already published and it has not been our intention to compete with these. Rather our object has been to explain as clearly as possible the basic phenomena and concepts of superconductivity in a manner which will be understood by those with no previous knowledge of superconductivity and only a modest acquaintance with solid-state physics.

In this book we have concentrated on the physics of superconductivity and, though we occasionally mention applications, we have not treated these in any detail. We hope, nevertheless, that this book will be useful both to "pure" physicists and to those interested in practical applications. With this in mind, we have used the rationalized MKS system throughout. Many of those interested in practical applications will be engineers brought up on the MKS system and, furthermore, MKS units are increasingly used in physics teaching. It seemed to us that someone should take the plunge and write a book on superconductivity using the MKS system. The MKS approach has involved us in some thought about the meaning of B and H in superconductors, a point which is discussed in Appendix A.

We have, of course, drawn on other texts on the subject and, in particular, we have made frequent use of Shoenberg's classic monograph *Superconductivity*. Because our book is an introduction, we have not attempted to include a complete list of references, but have referred to some of the key papers where this seemed appropriate.

We are grateful to Professor G. Rickayzen and to Mr. K. E. Osborne for discussing the contents of Chapter 9 and 11, and to those authors who allowed us to use the original copies of figures from their publications. We must also thank Mrs. Shirley Breen, who typed the manuscript and who, like Maxwell's daemon, made order where none was before. Our wives and children deserve gratitude for their forebearance during the time this book was being written.

University of Manchester   E.H.R.
  Institute of Science and Technology   A.C.R.-I.
    July 1968

SYMBOLS

THROUGHOUT this book there is employed, as far as possible, a consistent set of symbols. The following list includes definitions of most of the symbols used.

a	half-thickness of a slab
a_l	coefficient of wavefunction $\phi(\mathbf{p}_l\!\uparrow, -\mathbf{p}_l\!\downarrow)$ in pair wavefunction
$\mathbf{A}$	magnetic vector potential, defined by $\mathbf{B} = \operatorname{curl} \mathbf{A}$
$\mathscr{A}$	area
B	magnetic flux density
B_a	flux density of applied magnetic field (§ 2.1)
B_c	critical flux density ($= \mu_0 H_c$)
C_n	specific heat of normal phase
C_s	specific heat of superconducting phase
C_{el}	electronic contribution to specific heat
C_{latt}	lattice contribution to specific heat
d	thickness of a plate
e	electronic charge
E	energy *or* electric field strength
E_g	energy gap of superconductor ($= 2\Delta$)
g_n	Gibbs free energy per unit volume of normal phase
g_s	Gibbs free energy per unit volume of superconducting phase
G_n	Gibbs free energy of specimen in normal state
G_s	Gibbs free energy of specimen in superconducting state
$\hbar$	Planck's constant $\div\, 2\pi$
h_l	probability that pair state $(\mathbf{p}_l\!\uparrow, -\mathbf{p}_l\!\downarrow)$ is occupied in BCS ground state
H	magnetic field strength
H_a	applied magnetic field strength (see p. 16)
H_i	field strength inside specimen *or* field strength due to transport current
H_c	thermodynamic critical magnetic field strength

H_c'	enhanced critical magnetic field of a thin specimen
H_0	critical magnetic field strength at 0°K
H_{c1}	lower critical field of type-II superconductor
H_{c2}	upper critical field of type-II superconductor
i	current
i_c	critical current
i_s	supercurrent, current of electron-pairs
I	magnetization (magnetic moment per unit volume) or current
$\mathscr{I}$	current or current per unit width
j	surface current density per unit length
$\mathscr{J}$	volume current density per unit area (= magnitude of current density vector **J**)
$\mathscr{J}_c$	critical current density
$\mathscr{J}_n$	current density due to normal electrons
$\mathscr{J}_s$	current density due to superconducting electrons
k	Boltzmann's constant or effective susceptibility of plate [eqn. (8.4)]
l_e	electron mean free-path in normal state
L	self-inductance or latent heat
m	electronic mass or number of turns/unit length of solenoid
M	mutual inductance or total magnetic moment of specimen
n	demagnetizing factor [defined by eqn. (6.2)] or any integer
n_s	density of superelectrons
$\mathscr{N}(\varepsilon)$	density of states for electrons with kinetic energy ε
p	momentum of an electron or pressure
p_F	Fermi momentum [$= \sqrt{(2m\varepsilon_F)}$]
P	total momentum of a Cooper pair
q	electric charge or momentum of a phonon
R_n	resistance of specimen in normal state
R'	flow resistance of a type-II superconductor in the mixed state
s	entropy density or velocity of sound
T_c	transition temperature in zero field (or critical temperature)
u	internal energy per unit volume
$\mathbf{v}_s$	velocity of superelectrons
V	volume or matrix element of scattering interaction or voltage difference
W	kinetic energy
x_n	thickness of normal lamina in intermediate state
x_s	thickness of superconducting lamina in intermediate state
α	surface energy per unit area or a parameter $= m/\mu_0 n_s e^2$ (Chap. 3)

SYMBOLS

γ	Sommerfield specific heat constant [eqns. (5.6) and (9.11)]
Δ	surface energy parameter defined by $\alpha = \frac{1}{2}\mu_0 H_c^2 \Delta$ or a parameter defined by $\Delta = 2h\nu_L \exp\left[-\{\mathcal{N}(\varepsilon_F)V\}^{-1}\right]$, equal to half the energy gap
$\Delta(0)$	half energy gap at 0°K
ε	kinetic energy of an electron ($= p^2/2m$)
ε_F	Fermi energy
ζ	attenuation length for tunnelling wave function
η	fraction of body in normal state ($= x_n/(x_n + x_s)$) or viscosity coefficient for flux flow (Chap. 13)
ξ	coherence length (§ 6.9)
ξ_0	coherence length in pure superconductor
θ	angle or Debye temperature
$\varkappa$	Ginzburg–Landau parameter (§ 12.3)
λ	penetration depth [defined by eqn. (2.2)] or wavelength
λ_0	penetration depth at 0°K
λ_L	London penetration depth [defined by eqn. (3.13)]
μ_0	permeability of free space ($4\pi \times 10^{-7}$ Hm^{-1})
μ_r	relative permeability
ν	frequency
ν_0	threshold frequency for onset of absorption of electromagnetic radiation
ν_q	phonon frequency
ν_L	an "average" phonon frequency
ρ	electrical resistivity
ρ'	flow resistivity in mixed state
ρ_0	resistivity at 0°K
σ	electrical conductivity
χ	magnetic susceptibility
$\psi(x, y, z, \mathbf{p})$ or $\psi(\mathbf{p})$	one-electron wavefunction for non-interacting electron of momentum $\mathbf{p}$
ϕ	phase
$\Delta\phi$	phase difference
$\phi(x_1, y_1, z_1, \mathbf{p}_1, x_2, y_2, z_2, \mathbf{p}_2)$ or $\phi(\mathbf{p}_1, \mathbf{p}_2)$	two-electron wavefunction for non-interacting electrons with momenta $\mathbf{p}_1$ and $\mathbf{p}_2$
Φ	magnetic flux
Φ'	fluxoid [eqn. (11.5)]
Φ_0	fluxon [eqn. (11.8)]
$\Phi(x_1, y_1, z_1, x_2, y_2, z_2)$ or $\Phi(\mathbf{r}_1, \mathbf{r}_2)$	two-electron wavefunction for a pair of interacting electrons, defined by eqn. (9.4)

Φ_P	wavefunction of Cooper pair with total momentum P
Ψ	Ginzburg–Landau effective wavefunction (§ 8.5)
$\Psi_G(\mathbf{r}_1, \mathbf{r}_2, \ldots, \mathbf{r}_{n_s})$	many-electron wavefunction describing BCS ground state [eqn. (9.6)] and formally identical with the Ginzburg–Landau Ψ
Ψ_p	electron-pair wave function

INTRODUCTION

SUPERCONDUCTIVITY is the name given to a remarkable combination of electric and magnetic properties which appears in certain metals when they are cooled to extremely low temperatures. Such very low temperatures first became available in 1908 when Kamerlingh Onnes at the University of Leiden succeeded in liquefying helium, and by its use was able to obtain temperatures down to about 1°K.

One of the first investigations which Onnes carried out in the newly available low-temperature range was a study of the variation of the electrical resistance of metals with temperature. It had been known for many years that the resistance of metals falls when they are cooled below room temperature, but it was not known what limiting value the resistance would approach if the temperature were reduced towards 0°K. Onnes, experimenting with platinum, found that, when cooled, its resistance fell to a low value which depended on the purity of the specimen. At that time the purest available metal was mercury and, in an attempt to discover the behaviour of a very pure metal, Onnes measured the resistance of pure mercury. He found that at very low temperatures the resistance became immeasurably small, which was not surprising, but he soon discovered (1911) that the manner in which the resistance disappeared was completely unexpected. Instead of the resistance falling smoothly as the temperature was reduced towards 0°K, the resistance fell sharply at about 4°K, and below this temperature the mercury exhibited no resistance whatsoever. Furthermore, this sudden transition to a state of no resistance was not confined to the pure metal but occurred even if the mercury was quite impure. Onnes recognized that below 4°K mercury passes into a new state with electrical properties quite unlike those previously known, and this new state was called the "superconducting state".

It was later discovered that superconductivity could be destroyed (i.e. electrical resistance restored) if a sufficiently strong magnetic field were applied, and subsequently it was found that a metal in the superconducting state has very extraordinary magnetic properties, quite unlike those known at ordinary temperatures.

Up to the present time about half of the metallic elements and also a number of alloys have been found to become superconducting at low temperatures. Those metals which exhibit superconductivity when sufficiently cooled are called superconductors. For many years it was thought that all superconductors behaved according to a basically similar pattern. However, it is now realized that there are two kinds of superconductor, which are known as type-I and type-II. Most of those elements which are superconductors exhibit type-I superconductivity, whereas alloys generally exhibit type-II superconductivity. The two types have many properties in common but show considerable differences in their magnetic behaviour. These differences are sufficient for us to treat the two types separately. The first part of this book deals with type-I superconductors and the second part with type-II superconductors.

PART I

TYPE-I SUPERCONDUCTORS

CHAPTER 1

ZERO RESISTANCE

THE ELECTRICAL resistivity of all metals and alloys decreases when they are cooled. To understand why this should be, we must consider what causes a conductor to have resistance. The current in a conductor is carried by "conduction electrons" which are free to move through the material. Electrons have, of course, a wave-like nature, and an electron travelling through a metal can be represented by a plane wave progressing in the same direction. A metal has a crystalline structure with the atoms lying on a regular repetitive lattice, and it is a property of a plane wave that it can pass through a perfectly periodic structure without being scattered into other directions. Hence an electron is able to pass through a perfect crystal without any loss of momentum in its original direction. In other words, if in a perfect crystal we start a current flowing (which is equivalent to giving the conduction electrons a net momentum in the direction of the current) the current will experience no resistance. However, any fault in the periodicity of the crystal will scatter the electron wave and introduce some resistance. There are two effects which can spoil the perfect periodicity of a crystal lattice and so introduce resistance. At temperatures above absolute zero the atoms are vibrating and will be displaced by various amounts from their equilibrium positions; furthermore, foreign atoms or other defects randomly distributed can interrupt the perfect periodicity. Both the thermal vibrations and any impurities or imperfections scatter the moving conduction electrons and give rise to electrical resistance.

We can now see why the electrical resistivity decreases when a metal or alloy is cooled. When the temperature is lowered, the thermal vibrations of the atoms decrease and the conduction electrons are less frequently scattered. The decrease of resistance is linear down to a temperature equal to about one-third of the characteristic Debye temperature of the material, but below this the resistance decreases progressively less rapidly as the temperature falls (Fig. 1.1). For a perfectly pure metal, where the electron motion is impeded only by the thermal vibrations of the lattice, the resistivity should approach zero as the temperature is reduced towards 0°K. This zero resistance which a hypothetical "perfect" specimen would acquire if it could be cooled to absolute zero, is *not*, however, the phenomenon of superconductivity. Any real specimen of metal cannot be perfectly pure and will contain some impurities. Therefore the electrons, in addition to being scattered by thermal vibrations of the lattice atoms, are scattered by the impurities, and this impurity scattering is more or less independent of temperature. As a result, there is a certain "residual resistivity" (ρ_0, Fig. 1.1) which remains even at the lowest temperatures. The more impure the metal, the larger will be its residual resistivity.

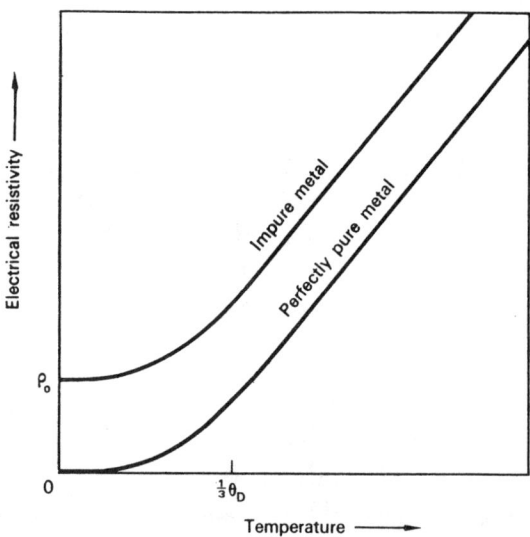

FIG. 1.1. Variation of resistance of metals with temperature.

Certain metals, however, show a very remarkable behaviour; when they are cooled their electrical resistance decreases in the usual way, but on reaching a temperature a few degrees above absolute zero they suddenly lose all trace of electrical resistance (Fig. 1.2). They are then said to have passed into the *superconducting* state.† The transformation

† In this book we use the term *superconductor* for a material which shows superconductivity if cooled. We use the adjective *superconducting* to describe it when it is exhibiting superconductivity, and *normal* when it is not exhibiting superconductivity (e.g. when above its transition temperature).

to the superconducting state may occur even if the metal is so impure that it would otherwise have had a large residual resistivity.

1.1. Superconducting Transition Temperature

The temperature at which a superconductor loses resistance is called its superconducting *transition temperature* or *critical temperature*; this temperature, written T_c, is different for each metal. Table 1.1 shows the transition temperatures for metallic elements. In general the transition temperature is not very sensitive to small amounts of impurity, though

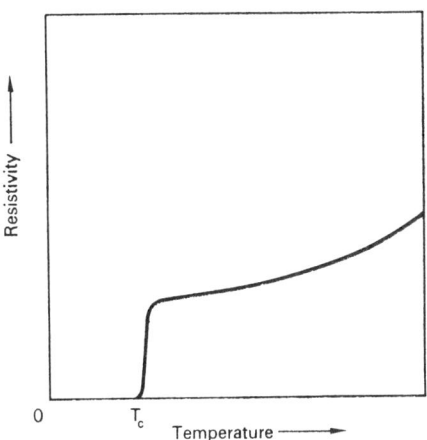

FIG. 1.2 Loss of resistance of a superconductor at low temperatures.

magnetic impurities tend to lower the transition temperature. (We shall see in Chapter 9 that ferromagnetism, in which the spins of electrons are aligned parallel to each other, is incompatible with superconductivity.) The superconductivity of a few metals, such as iridium and molybdenum, which in the pure state have very low transition temperatures, may be destroyed by the presence of minute quantities of magnetic impurities. Such elements, therefore, only exhibit superconductivity if they are extremely pure, and specimens of these metals of normal commercial purity are not superconductors. Not all pure metals have been found to be superconductors; for example, copper, iron and sodium have not shown superconductivity down to the lowest temperature to which they have so far been cooled. Of course, experiments at even lower temperatures may reveal new superconductors, but there is no fundamental reason why all metals should show superconductivity, even at absolute zero. Nevertheless, it should be realized that superconductivity is not a rare phenomenon; about half the metallic elements are known to be superconductors and, in addition, a large number of alloys are superconductors. It is possible for an alloy to be a superconductor, even if it is composed of two metals which are not themselves superconductors (e.g. Bi–Pd). Superconductivity can be shown by conductors which are not metals in the ordinary sense; for example, the semiconducting mixed oxide of barium, lead and bismuth is a superconductor, and the conducting polymer, polysulphur nitride $(SN)_x$, has been found to become superconducting at about $0\cdot3°K$.

TABLE 1.1. THE SUPERCONDUCTING ELEMENTS
T_c is the superconducting transition temperature
H_0 is the critical magnetic field at $0°K$ (see Chapter 4)

	$T_c(°K)$	H_0 (Amp m^{-1})	H_0 (gauss)
Aluminium	1·2	$0·79 \times 10^4$	99
Cadmium	0·52	$0·22 \times 10^4$	30
Gallium	1·1	$0·41 \times 10^4$	51
Indium	3·4	$2·2 \times 10^4$	276
Iridium	0·11	$0·13 \times 10^4$	16
Lanthanum $\begin{cases} \alpha \\ \beta \end{cases}$	4·8 / 4·9		
Lead	7·2	$6·4 \times 10^4$	803
Lutecium	0·1	$2·8 \times 10^4$	350
Mercury $\begin{cases} \alpha \\ \beta \end{cases}$	4·2 / 4·0	$3·3 \times 10^4$ / $2·7 \times 10^4$	413 / 340
Molybdenum	0·9		
Niobium	9·3	Type-II (see Chap. 12)	
Osmium	0·7	$\sim 0·5 \times 10^4$	~ 63
Rhenium	1·7	$1·6 \times 10^4$	201
Ruthenium	0·5	$0·53 \times 10^4$	66
Tantalum	4·5	$6·6 \times 10^4$	830
Technetium	7·9	Type-II (see Chap. 12)	
Thalium	2·4	$1·4 \times 10^4$	171
Thorium	1·4	$1·3 \times 10^4$	162
Tin	3·7	$2·4 \times 10^4$	306
Titanium	0·4		
Tungsten	0·016	$0·0096 \times 10^4$	1·2
Uranium $\begin{cases} \alpha \\ \beta \end{cases}$	0·6 / 1·8		
Vanadium	5·4	Type-II (see Chap. 12)	
Zinc	0·9	$0·42 \times 10^4$	53
Zirconium	0·8	$0·37 \times 10^4$	47

Niobium is the metallic element with the highest transition temperature (9·3°K), but some alloys and metallic compounds remain superconducting up to even higher temperatures (Table 1.2). For example, Nb$_3$Ge has a transition temperature of about 23°K. These alloys with relatively high transition temperatures are of great importance in the engineering applications of superconductivity.

TABLE 1.2. SUPERCONDUCTING TRANSITION TEMPERATURES OF SOME ALLOYS AND METALLIC COMPOUNDS COMPARED WITH THEIR CONSTITUENT ELEMENTS

	Ta–Nb	Pb–Bi	3Nb–Zr	Nb$_3$Sn	Nb$_3$Ge
T_c (°K)	6·3	8	11	18	23

	Nb	Pb	Ta	Sn	Zr	Bi	Ge
T_c (°K)	9·3	7·2	4·5	3·7	0·8	not s/c	not s/c

On cooling, the transition to the superconducting state may be extremely sharp if the specimen is pure and physically perfect. For example, in a good gallium specimen, the transition has been observed to occur within a temperature range of 10^{-5} degrees. If, however, the specimen is impure or has a disturbed crystal structure, the transition may be considerably broadened. Figure 1.3 shows the transition in pure and impure tin specimens.

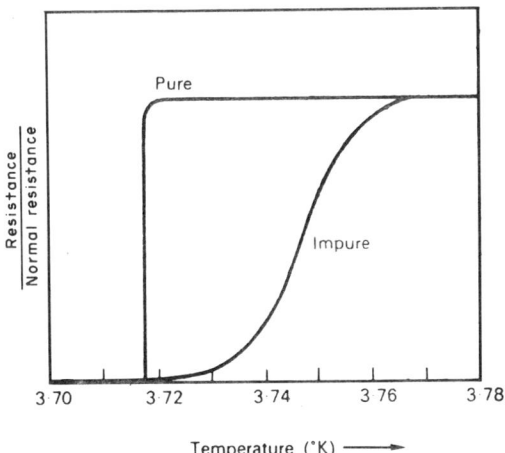

FIG. 1.3. Superconducting transition in tin.

1.2 Zero Resistance

Even when the transition is spread over a considerable temperature range the resistance still seems to disappear completely below a certain temperature. We naturally ask whether in the superconducting state the resistance has indeed become zero or whether it has merely fallen to a very small value. Of course, it can never be proved by experiment that the resistance is in fact zero; the resistance of any specimen may always be just less than the sensitivity of our apparatus allows us to detect. However, no experiment has been able to detect any resistance in the superconducting state. We may look for resistance quite simply by passing a current through a wire of superconductor and seeing if any voltage is recorded by a sensitive voltmeter connected across the ends of the wire. A more sensitive test, however, is to start a current flowing round a closed superconducting ring and then see whether there is any decay in the current after a long period of time. Suppose the self-inductance of the ring is L; then, if at time $t = 0$ we start a current $i(0)$ flowing round the ring (for ways of doing this see § 1.3), at a later time t the current will have decayed to

$$i(t) = i(0)e^{-(R/L)t}, \tag{1.1}$$

where R is the resistance of the ring. We can measure the magnetic field that the circulating current produces and see if this decays with time. The measurement of the magnetic field does not draw energy from the circuit, and we should be able to observe whether the current circulates indefinitely. Gallop has been able to show from the lack of decay of a current circulating round a closed loop of superconducting wire that the resistivity of the superconducting metal was less than 10^{-26} ohm-metres (i.e. less than 10^{-18} the resistivity of copper at room temperature). It seems, therefore, that we are justified in treating the resistance of a superconducting metal as zero.

1.3. The Resistanceless Circuit

A closed circuit, such as a ring, formed of superconducting metal has an important and useful property resulting from its zero resistance. *The total magnetic flux threading a closed resistanceless circuit cannot change so long as the circuit remains resistanceless.* Suppose as in Fig. 1.4a, a ring of metal is cooled below its transition temperature in an applied field of uniform flux density B_a. If the area enclosed by the ring is $\mathscr{A}$, an amount of flux $\Phi = \mathscr{A} B_a$ will thread the ring. Suppose the applied field is now changed to a new value. By Lenz's law, when the field is changing currents are induced and circulate round the ring in such a direction as to create a flux inside the ring which tends to cancel the flux change due to the alteration in the applied field. While the field is changing there is an e.m.f., $-\mathscr{A} dB_a/dt$, and an induced current i given by

$$-\mathscr{A}\frac{dB_a}{dt} = Ri + L\frac{di}{dt},$$

where R and L are the total resistance and inductance of the circuit. In a normal resistive circuit the induced currents quickly die away and the flux threading the ring acquires the new value. In a superconducting circuit, however, $R = 0$ and

$$-\mathscr{A}\frac{dB_a}{dt} = L\frac{di}{dt},$$

so that $\qquad Li + \mathscr{A} B_a = \text{constant}.$ (1.2)

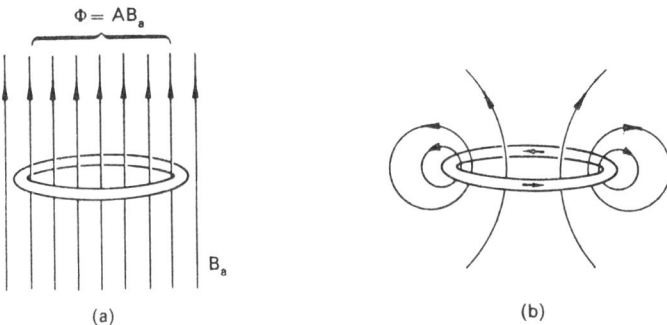

FIG. 1.4. Resistanceless circuit.

But $Li + \mathscr{A} B_a$ is the total magnetic flux threading the circuit; we have therefore demonstrated that the total flux threading a resistanceless circuit cannot change. If the applied magnetic field strength is changed an induced current is set up of such a magnitude that it creates a flux which exactly compensates the change in the flux from the applied magnetic field. Since the circuit is resistanceless the induced current flows for ever and the original amount of flux is maintained indefinitely. Even if the external field is reduced to zero, as in Fig. 1.4b, the internal flux will be maintained by the induced circulating current.

This property can be made use of when solenoids wound from superconducting wire are employed to generate magnetic fields. In Fig. 1.5 current to the refrigerated superconducting solenoid S is derived from the d.c. power supply P. Once the current has been adjusted by the rheostat R to a value which gives the desired magnetic field strength, the superconducting switch XY can be closed. XY and S now form a closed resistanceless circuit in which the flux must remain constant. Hence the field strength generated by S does not vary in time and we can, if we wish, disconnect the power supply and the field will be maintained by the current i flowing round the resistanceless circuit XYS. A superconducting solenoid operating in this fashion is said to be in the *persistent mode*.

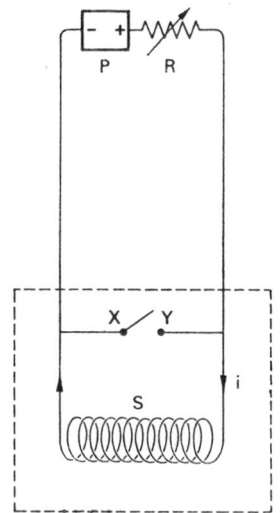

FIG. 1.5. Superconducting solenoid.

Note that, although the *total* amount of flux enclosed in a resistanceless circuit remains constant, there can be a change in the flux density **B** at any point due to a redistribution of the flux within the circuit. Thus in Fig. 1.4b the flux density has become stronger near the wire and weaker in the centre of the enclosed space compared to the uniform distribution in Fig. 1.4a. In both cases, however, the total flux $(= \iint_{\mathscr{A}} \mathbf{B} \cdot d\mathscr{A})$ is the same.

We have seen that if a closed circuit of superconductor is cooled below its transition temperature while in an applied magnetic field, the flux it encloses remains constant in spite of any changes in the applied field. On the other hand, if the circuit is cooled in the absence of an applied magnetic field, so that there is initially no flux inside, and an external magnetic field is subsequently applied, the net internal flux remains zero in spite of the presence of the external field. This property enables us to use hollow superconducting cylinders to shield enclosures from external magnetic fields. The shielding is only perfect for the case of a long hollow cylinder, in which case the induced currents generate a uniform compensating flux density throughout its interior. For other configurations, such as a short ring, it is only the *total* flux which is maintained at zero and the local magnetic flux density generated by the induced current will not be uniform within the ring. Hence the flux density due to the persistent currents will in some places be stronger and in other places weaker than that of an applied field, and there will not everywhere be exact cancellation. In other words, though $\iint_\mathscr{A} \mathbf{B} \cdot d\mathscr{A} = 0$, $\mathbf{B}$ itself is not necessarily

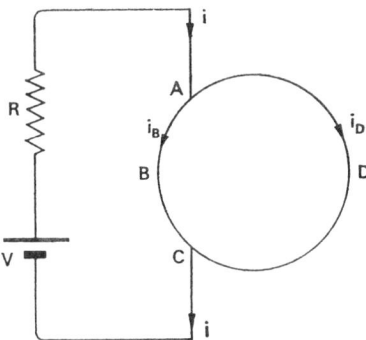

FIG. 1.6. Division of current between two parallel paths.

everywhere equal to zero. In practice, however, superconducting shielding can be used to give very good screening against magnetic fields.

We now consider what determines the distribution of currents in a network of resistanceless conductors. Consider, for example, the simple circuit of Fig. 1.6. If the ring $ABCD$ is resistanceless, how will the current i divide between branches B and D? Clearly Kirchhoff's laws are of no help, because both paths have zero resistance and the second law will be obeyed for all possible divisions of the current i. However, though the two branches have no resistance, they do contribute inductance to the circuit. We shall now show that the division of current is determined by these inductances. The voltage difference between A and C equals

$$L_B \frac{di_B}{dt} + M_{BD} \frac{di_D}{dt} = L_D \frac{di_D}{dt} + M_{BD} \frac{di_B}{dt},$$

where L_B and L_D are the inductances of the branches B and D, and M_{BD} is the mutual inductance between them. Rearrangement gives

$$(L_B - M_{BD}) \frac{di_B}{dt} = (L_D - M_{BD}) \frac{di_D}{dt},$$

and integrating this we obtain

$$(L_B - M_{BD}) i_B = (L_D - M_{BD}) i_D + \text{constant}.$$

If $i_B = 0 = i_D$ at $t = 0$, then the constant $= 0$, and we have

$$\frac{i_B}{i_D} = \frac{L_D - M_{BD}}{L_B - M_{BD}},$$

and it can be seen that the division of the current is controlled by the inductances of the paths. It often happens that the magnetic coupling between two parallel paths is small so that their mutual inductance can be neglected. Under this circumstance we can deduce the rule that in resistanceless networks *the currents carried in parallel paths are inversely proportional to the self-inductances of those paths*.

1.4. A.C. Resistivity

The fact that a superconducting metal has no resistance means, of course, that there is no voltage drop along the metal when a current is passed through it, and no power is generated by the passage of the current. This, however, is only strictly true for a direct current of constant value. If the current is changing an electric field is developed and some power is dissipated. To understand the reason for this we must first discuss briefly some aspects of the behaviour of conduction electrons in superconductors.

Many of the properties of superconductors can be explained if it is supposed that below the transition temperature the conduction electrons divide into two classes, some behaving as "superelectrons" which can pass through the metal without resistance (i.e. suffering no collisions), the remainder behaving as "normal" electrons which can be scattered and so experience resistance just like conduction electrons in a normal metal. The fraction of superelectrons appears to decrease as the temperature is raised towards the transition temperature. At $0°K$ all conduction electrons behave like superelectrons, but, if the temperature is raised, a few begin to behave as normal electrons, and on further heating the proportion of normal electrons increases. Eventually, at the transition temperature, all the electrons have become normal electrons and the metal loses its superconductive properties. Hence a superconductor below its transition temperature appears to be permeated by two electron fluids, one of normal electrons and one of superelectrons. The relative electron density in the two fluids depends on the temperature. This "two-fluid model" is suggested by thermodynamic arguments based on the results of specific heat and similar measurements on superconductors, which will be discussed in Chapter 5.

In a superconducting metal the current can in general be carried by both the normal and superelectrons. However, in the special case of a *constant* direct current all the current is carried by the superelectrons. We can see that this will be so by noting that, if the current is to remain constant, there must be no electric field in the metal, otherwise the superelectrons would be accelerated continuously in this field and the current would increase indefinitely. If there is no field there is nothing to drive the normal electrons and so there is no normal current. We see, therefore, that for a constant value of total current all the current is carried by the superelectrons. A superconducting metal is like two conductors in parallel, one having a normal resistance and the other zero resistance. We can say that the superelectrons "short circuit" the normal electrons. To put this another way, if we suddenly apply a voltage source, such as a battery, across a superconductor, the current tends to rise to infinity but is in fact limited by the internal resistance of the source. While the current is changing, an electric field must be present to accelerate the electrons. Electrons do, however, have a small inertial mass and so the supercurrent does not rise instantaneously but only at the rate at which the electrons accelerate in the electric field. If we apply an *alternating* field, the supercurrent will therefore lag behind the field because of the inertia of the superelectrons. Hence the superelectrons present an inductive impedance† and, because there now is an electric field present, some of the current will be carried by the normal electrons. The current is not, therefore, carried entirely by the superelectrons as in the d.c. case. Of course, the normal electrons also have an inertial mass but their resulting inductive reactance is completely swamped by the

---

† This inherent inductive impedance is, of course, quite distinct from, and is additional to, the ordinary inductance of the conductor due to its geometry.

resistance resulting from their being scattered in the metal. We can, in fact, represent the bulk properties of a superconducting metal by a perfect inductance in parallel with a resistance.

The fraction of the current diverted through the normal electrons dissipates power in the usual way. The mass of electrons is, however, extremely small, so the inductance due to their inertia is also extremely small. The inductance in henrys of a typical superconductor due to the inertia of its superelectrons is only about 10^{-12} of its normal resistance in ohms, so at 1000 Hz, for example, only about 10^{-8} of the total current is carried by the normal electrons and there is only a minute dissipation of power. Nevertheless, this contrasts with the absolutely zero resistance in the d.c. case.

If the frequency of an applied field is sufficiently high, however, a superconducting metal responds in the same way as a normal metal. This is because, as we shall see in Chapter 9, superelectrons are in a lower energy state than normal electrons, but, if the frequency of the applied field is high enough, the photons of the electromagnetic field have enough energy to excite superelectrons into the higher state where they behave as normal electrons. This happens for frequencies greater than about 10^{11} Hz (i.e. greater than the frequency of very long wave infrared). The behaviour of a superconductor at optical frequencies is therefore no different from that of a normal metal and there is, for example, no change in the visual appearance of a superconductor as it is cooled below its transition temperature.

It is tempting to suppose that the superelectrons in a superconductor behave like electrons in a vacuum. The electrons in the beam of a cathode ray tube, for example, are resistanceless in the sense that they flow without undergoing any collisions. There is, however, a significant difference between the two cases. It is possible to maintain a potential drop along an electron beam while the current remains at a constant value. This is because, though the current must be the same all along the path of the beam, the electron density need not remain constant. Hence the electrons accelerate from the cathode towards the anode and the electron density is relatively high near the cathode and decreases as the anode is approached. However, the product of the electron density and electron velocity, i.e. the current, remains constant along the beam. The fact that the electrons are able to accelerate allows us to maintain the electric field. In a superconductor, however, conditions are different. The metal must remain everywhere electrically neutral and, since the positive metal ions are fixed in the crystal, the electron density cannot vary through the material. Hence for a constant current to be maintained through the metal, the velocity of all the electrons along the current path must be the same. The electrons, therefore, do not accelerate and an electric field cannot exist in the metal.

CHAPTER 2

PERFECT DIAMAGNETISM

2.1. Magnetic Properties of a Perfect Conductor

WE HAVE seen in the previous chapter that a superconductor below its transition temperature appears to have no resistance. Let us now try to deduce the magnetic properties of such a resistanceless conductor.

Suppose that we cool a specimen which, below its transition temperature, becomes perfectly conducting. The resistance around an imaginary closed path within the metal is zero; and therefore, as shown in the previous chapter, the amount of magnetic flux enclosed within this path cannot change. This is true for *any* such imaginary circuit and this can only be so if the flux density at every point within the metal does not vary with time, i.e.

$$\dot{\mathbf{B}} = 0.$$

Consequently the flux distribution in the metal must remain as it was when the metal became resistanceless.

Consider now the behaviour of a perfect conductor under various circumstances. Suppose that a specimen loses its resistance in the absence of any magnetic field and that a magnetic field is then subsequently applied. Because the flux density in the metal cannot change, it must remain zero even after the application of the magnetic field. In fact the application of the magnetic field induces resistanceless currents which circulate on the surface of the specimen in such a manner as to create a magnetic flux density which everywhere inside the metal is exactly equal and opposite to the flux density of the applied magnetic field.† Because

---

† We must define here exactly what we mean by an "applied magnetic field". An "applied" field is a field generated by some agency (solenoid, permanent magnet, etc.) external to a specimen. Its strength H_a and flux density B_a are those which would be measured if the specimen were not there. In the case of a uniform applied field the strength and flux density are the same as those measured far away from the specimen, i.e. where any perturbing effects due to the magnetic properties of the specimen are negligible.

these currents do not die away, the net flux density inside the material remains at zero. This is illustrated in Fig. 2.1a: the surface currents i generate a flux density B_i that exactly cancels the flux density B_a of the applied magnetic field everywhere inside the metal. These surface currents are often referred to as *screening currents*.

The flux density created by the persistent surface currents does not, of course, disappear at the boundary of the specimen, but the flux lines form continuous closed curves which return through the space outside (Fig. 2.1a). Though the density of this flux everywhere *inside* the specimen is equal and opposite to the flux from the applied field, this is not so *outside* the specimen. The net distribution of flux resulting from the superposition of the flux from the specimen and that from the applied field is shown in Fig. 2.1b. The pattern is as though the sample

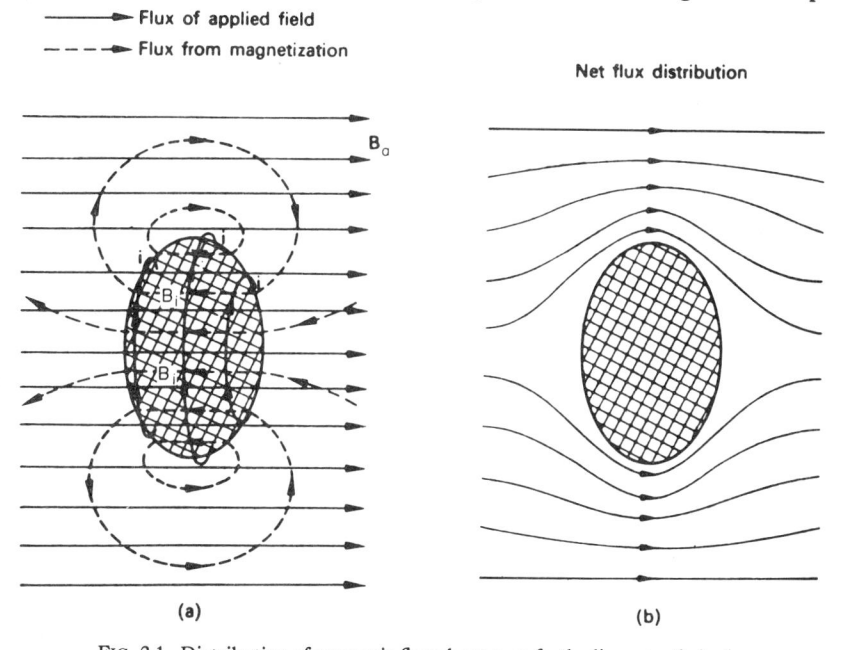

FIG. 2.1. Distribution of magnetic flux about a perfectly diamagnetic body.

had prevented entry into it of the flux of the applied field. A sample in which there is no net flux density when a magnetic field is applied is said to exhibit *perfect diamagnetism*. If we now reduce the applied magnetic field to zero the specimen is left in its original unmagnetized condition. The above sequence of events is illustrated in Fig. 2.2 a–d.

Let us now consider a different sequence of events. Suppose that the magnetic field B_a is applied to the specimen while it is above its transi-

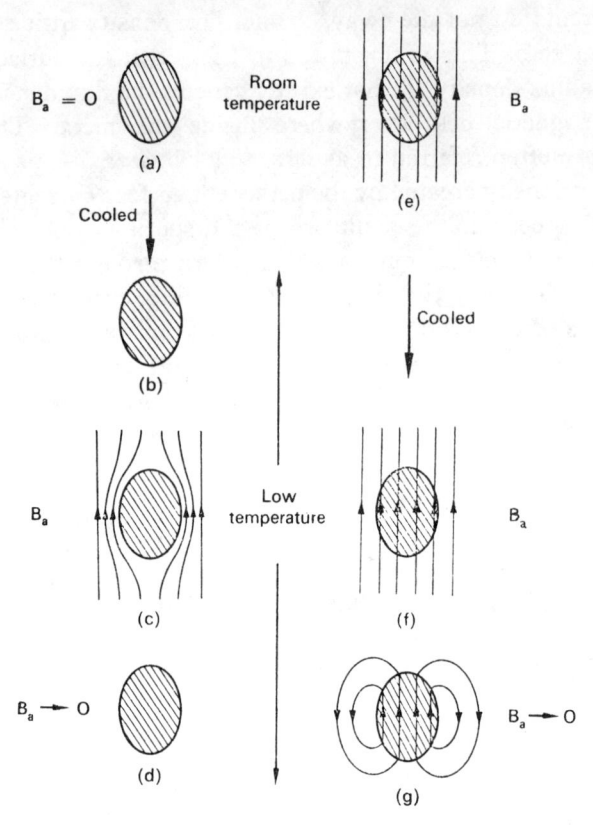

FIG. 2.2. Magnetic behaviour of a "perfect" conductor. (a)–(b) Specimen becomes resistanceless in absence of field. (c) Magnetic field applied to resistanceless specimen. (d) Magnetic field removed.

(e)–(f) Specimen becomes resistanceless in applied magnetic field. (g) Applied magnetic field removed.

tion temperature (Fig. 2.2e). Most metals (other than the special ferromagnetics, iron, cobalt and nickel) have values of relative magnetic permeability very close to unity, and so the flux density inside is virtually the same as that of the applied field. The specimen is now cooled to a low temperature so that it loses its electrical resistance. This disappearance of resistance has no effect on the magnetization, and the flux distribution remains unaltered (Fig. 2.2f). We next reduce the applied field to zero. The flux density inside the perfectly conducting metal cannot change, and persistent currents are induced on the specimen, maintaining the flux inside, with the result that the specimen is left permanently magnetized (Fig. 2.2g).

It is important to notice that in (c) and (f) of Fig. 2.2 the sample is under the same conditions of temperature and applied magnetic field, and yet its state of magnetization is quite different in the two cases. Similarly (d) and (g) show different states of magnetization under identical external conditions. We see that the state of magnetization of a perfect conductor is not uniquely determined by the external conditions but depends on the sequence by which these conditions were arrived at.

2.2. Special Magnetic Behaviour of a Superconductor

2.2.1. Meissner effect

In the previous section we have deduced the magnetic behaviour of a resistanceless conductor by applying simple and well-known fundamental principles of electromagnetism, and for 22 years after the discovery of superconductivity it was assumed that the effect of a magnetic field on a superconductor would be as shown in Fig. 2.2. However, in 1933 Meissner and Ochsenfeld measured the flux distribution outside tin and lead specimens which had been cooled below their transition temperatures while in a magnetic field.† They found that the expected situation of Fig. 2.2f did not in fact occur, but that at their transition temperatures the specimens spontaneously became perfectly diamagnetic, cancelling all flux inside, as in Fig. 2.2c, even though they had been cooled in a magnetic field. This experiment was the first to demonstrate that superconductors are something more than materials which are perfectly conducting; they have an additional property that a merely resistanceless metal would not possess: *a metal in the superconducting state never allows a magnetic flux density to exist in its interior.* That is to say, inside a superconducting metal we always have

$$\mathbf{B} = 0,$$

whereas inside a merely resistanceless metal there may or may not be a flux density, depending on circumstance (Fig. 2.2). When a superconductor is cooled in a weak magnetic field, at the transition temperature persistent currents arise on the surface and circulate so as to cancel the flux density inside, in just the same way as when a magnetic field is applied

---

† The magnetic field used in experiments of this kind must not be too strong because, as we shall see in Chapter 4, a metal loses its superconducting properties if the applied magnetic field exceeds a certain strength.

after the metal has been cooled (Fig. 2.3). This effect, whereby a superconductor never has a flux density inside even when in an applied magnetic field, is called (with injustice to Ochsenfeld) *the Meissner effect*.

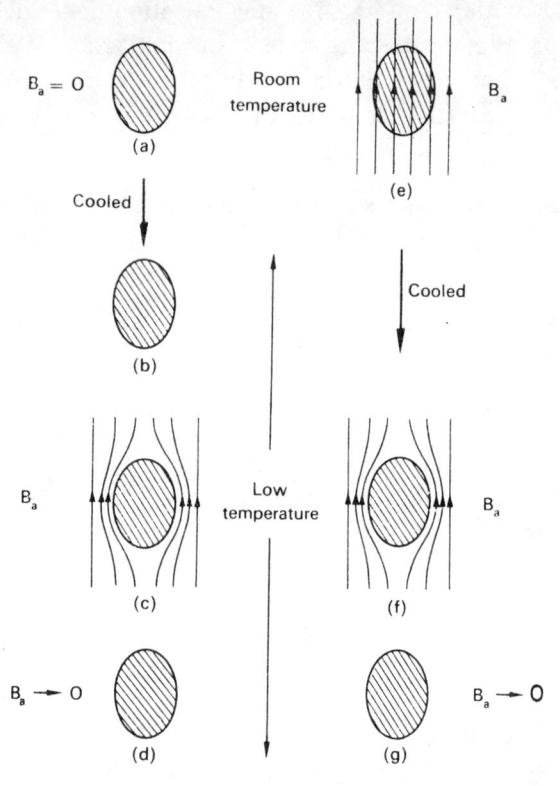

FIG. 2.3. Magnetic behaviour of a superconductor. (a)–(b) Specimen becomes resistanceless in absence of magnetic field. (c) Magnetic field applied to superconducting specimen. (d) Magnetic field removed.

(e)–(f) Specimen becomes superconducting in applied magnetic field. (g) Applied magnetic field removed.

For convenience we call a hypothetical metal which simply has no resistance and would behave as shown in Fig. 2.2 a "perfect conductor" in contrast to *super*conductors which in the superconducting state never permit a magnetic flux density to exist inside them (Fig. 2.3). The state of magnetization of a "perfect conductor" would depend on the order in which the final conditions of applied magnetic field and temperature were obtained, but the magnetization of a superconductor depends only on the actual values of the applied field and temperature and not on the way they were arrived at.

2.2.2. Permeability and susceptibility of a superconductor

Suppose a magnetic field of flux density B_a is applied to a superconductor.† In order that we may neglect demagnetizing effects (see Chapter 6) we consider a long superconducting rod with the field applied parallel to its length. An applied magnetic field of flux density B_a produces in the material a flux density equal to $\mu_r B_a$, where μ_r is the relative permeability of the material. Metals, other than ferromagnetics, have a relative permeability which is very close to unity, i.e. $\mu_r = 1$, so the flux density within them due to the applied magnetic field is equal to B_a. However, as we have seen, the total flux density inside a superconducting body is zero. This perfect diamagnetism arises because surface screening currents circulate so as to produce a flux density B_i which everywhere inside the metal exactly cancels the flux density due to the applied field; $B_i = -B_a$. A rod-shaped superconducting specimen therefore behaves like a long solenoid with circulating current that creates a flux density exactly equal in magnitude, but opposite in direction, to the flux density due to the applied magnetic field. To create a flux density of $-B_a$, the magnitude of the circulating surface current per unit length must, from the ordinary solenoid formula, be $|j| = B_a/\mu_0$. In other words $|j| = H_a$, where H_a is the applied field strength.

We can, however, describe the perfect diamagnetism in another way. Because we cannot actually observe the surface screening currents which arise when a magnetic field is applied, we could suppose that the perfect diamagnetism arises from some special bulk magnetic property of the superconducting metal, and we can describe the perfect diamagnetism simply by saying that for a superconducting metal $\mu_r = 0$, so that the flux density inside, $B = \mu_r B_a$, is zero. Here we do not consider the mechanism by which the diamagnetism arises; the effect of the screening currents is included in the statement $\mu_r = 0$. The strength H_a of the applied magnetic field is given by

$$H_a = \frac{B_a}{\mu_0},$$

and the flux density in a magnetic material is related to the strength of the applied field by

$$B = \mu_0 (H_a + I),$$

---

† As explained in Appendix A, we take the view that a magnetic field is best described by its flux density.

where I is the magnetization (often called the "intensity of magnetization") of the material. The magnetization of a superconductor, in which $B = 0$, must therefore be

$$I = -H_a,$$

and the magnetic susceptibility, i.e. the ratio of the magnetization to the field strength, must be

$$\chi = -1.$$

The two descriptions are entirely equivalent because, as shown in Appendix A, the magnitude of I is equal to the equivalent surface current density j.

We now summarize the two alternative ways of regarding the perfect diamagnetism.

(i) Screening-current diamagnetism

The material of the superconductor, like other metals, is non-magnetic, and an applied magnetic field produces a flux density B_a in the metal. However, screening currents generate an internal flux density which everywhere is exactly equal and opposite to this flux density and consequently the net flux density is zero.

(ii) Bulk diamagnetism

The material can be considered to have a relative permeability $\mu_r = 0$, so the flux density produced in it by an applied magnetic field is always zero. The material behaves as though in a magnetic field it acquires a negative bulk magnetization $I = -H_a$.

Appendix A shows that these two ways of regarding the perfect diamagnetism are entirely equivalent. We shall make use of both descriptions, the choice in each case depending on which is more convenient for the particular situation under discussion.

2.3. Surface Currents

The fact that a superconducting metal does not allow magnetic flux to exist in its interior has an important effect on any electric currents that flow along it; *currents cannot pass through the body of a superconducting metal, but flow only on the surface.* To see why this should be, let us take the first viewpoint of a superconductor expressed at (i) above and treat the material as having the same relative magnetic permeability as any ordinary metal, i.e. $\mu_r = 1$. At any point in a material of unit relative permeability the relation between the magnetic flux and current density is, from Maxwell's equation,

$$\text{curl } \mathbf{B} = \mu_0 \mathbf{J}. \tag{2.1}$$

If the metal is superconducting **B** is zero inside, and so curl **B** must also be zero inside.† It follows, therefore, from Maxwell's equation that, as a consequence of **B** being zero, the current density **J** must also be zero within the superconductor. There is, of course, no reason why **B** must be

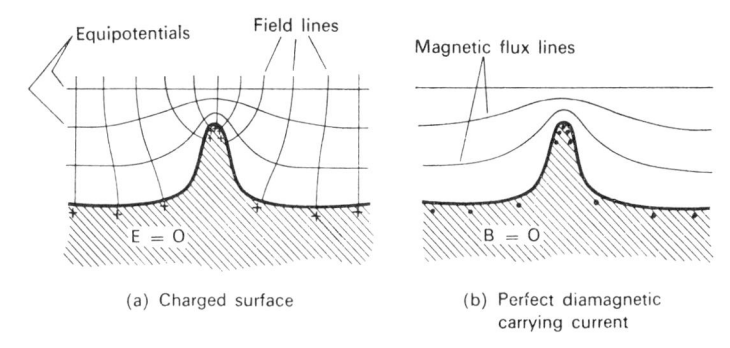

FIG. 2.4. Analogy between distribution of electrostatic charge and surface current.
+ + + electric charges
. . . current perpendicular to plane of page.

zero *outside* the superconductor so, if there is a current, it flows not through the metal but on the surface. This is true both of currents passed along the superconductor from some external source such as a battery (we call these "transport" currents), and of diamagnetic screening currents. Any transport current will flow all over the surface, creating a magnetic flux outside but not inside the conductor. If a magnetic field is applied, the diamagnetic screening currents which flow so as to cancel the flux density inside also circulate on the surface.

There is an interesting and useful analogy between the distribution of current on the surface of a superconducting metal and the distribution of electrostatic charge on a conducting body. Consider part of the surface of a charged conductor as shown in Fig. 2.4a. In the equilibrium state $E = 0$ inside the conductor but, if the body carries a surface charge, this charge will produce an electric field outside the conductor. The compo-

---

† As explained in Appendix A, we take the view that **J** affects **B** but not **H**. So the vanishing of **B** does not necessarily imply that **H** is zero.

nent of the electric field parallel to the surface, $E_\parallel$, is continuous across the surface and therefore, since $E = 0$ in the conductor, $E_\parallel$ must be zero just outside the surface. The electric field lines must consequently meet the conductor at right angles. The surface itself is an equipotential and the electric field lines are orthogonal to the equipotentials. It can be seen that the field lines are crowded together where the boundary has a high convex curvature, so the electric charge, being proportional to the normal component of the field, will be concentrated into these regions. Figure 2.4b shows a section of part of a superconducting metal carrying a current in a direction normal to the plane of the paper. Inside the perfectly diamagnetic material we have $B = 0$, but if current is flowing on the surface there will be a magnetic flux density outside. The component of B normal to the surface is continuous across the boundary, so just outside the surface the flux lines must be parallel to the surface. This flux density is proportional to the surface current density. In fact the external magnetic field due to the surface current has the same form as the equipotentials due to surface charges in Fig. 2.4a. The magnetic field lines are crowded together close to the region where the surface has high convex curvature, so the surface current density must be greatest at these regions. We might expect, therefore, that the distribution of surface current on a perfectly diamagnetic body has the same form as the distribution of electric charge on a charged conductor of the same shape, and a proper mathematical analysis shows that this is indeed so.

2.3.1. Hole through a superconductor

In later chapters we shall on several occasions need to consider the properties of a superconducting body with a hole right through it. Though the body of a superconductor is perfectly diamagnetic, flux can exist inside a hole.

Consider for example a long hollow body as shown in Fig. 2.5. This forms a closed circuit. We have already discussed the properties of closed resistanceless circuits in § 1.3, but, when the body is a superconductor, we must also take into account the perfect diamagnetism of the material itself.

First suppose that the body shown in Fig. 2.5 is cooled below its superconducting transition temperature in the absence of any applied magnetic field and that, after it has become superconducting, a magnetic field of flux density B_a is applied (Fig. 2.5a). The superconducting material is perfectly diamagnetic so there will be no flux density in the

PERFECT DIAMAGNETISM 25

body of the material. The perfect diamagnetism of the superconducting material is maintained by currents i_d which circulate on the outer surface so as to cancel the flux in the metal. However, the flux density generated by these diamagnetic screening currents also cancels the flux density due to the applied field in the hole, so in this case there is no flux density in the hole. In this situation the superconducting body is behaving no differently from a merely resistanceless body. In both cases the current induced on the outer surface by the application of the magnetic field cancels the flux inside.

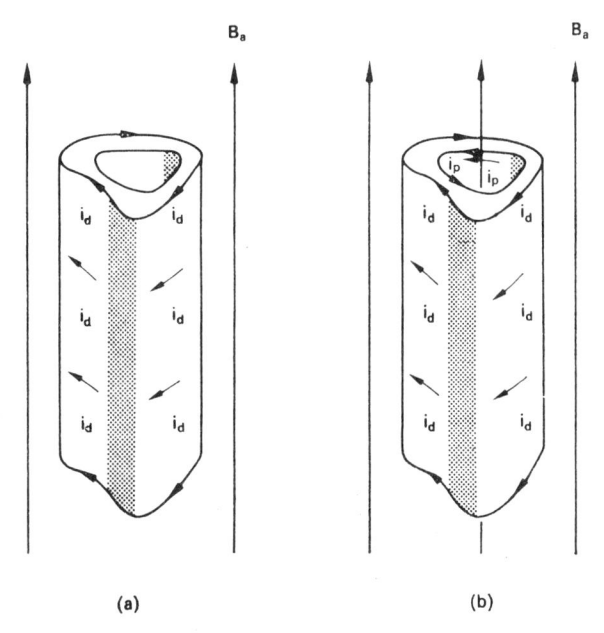

FIG. 2.5. Hollow superconductor. (a) Magnetic field applied when material is in the superconducting state. (b) Material becomes superconducting in an applied magnetic field.

Now consider a situation in which a superconductor behaves differently from a merely resistanceless body, or "perfect" conductor. Suppose the magnetic field is applied *before* the body is cooled below its transition temperature. Above the transition temperature the magnetic flux passes both through the body and the hole. In the case of the perfect conductor this flux distribution will not change when the body loses its resistance, and no currents will appear on the surface. A superconductor, however, behaves differently; below the transition temperature the material becomes perfectly diamagnetic but though there is no flux through the material, flux remains in the hole (Fig. 2.5b). Currents must circulate to maintain these differences in flux density. As we have just seen, the diamagnetic surface currents i_d which cancel the flux in the superconducting material would also cancel the flux in the hole, so if there is magnetic flux threading the hole, this must be generated by currents i_p circulating in the opposite ("paramagnetic") direction around its periphery. We have, therefore, the result that flux threading a hole or normal region through a superconductor is always associated with a current circulating around the boundary between this region and the superconductor.

Note that the net circulating current $i_p - i_d$ is just that which generates a flux density equal to the difference between the flux density in the hole and the flux density outside the superconducting body.

As we saw in Chapter 1, the flux threading any resistanceless circuit cannot change. Consequently once flux has been established through a hole in a superconducting body, this flux, and the circulating current i_p associated with it, will persist even if the applied magnetic field strength is changed or reduced to zero.

2.4. Penetration Depth

In § 2.3 we have seen that the perfect diamagnetism of a superconductor prevents electric currents flowing through the body of the material. On the other hand, currents cannot be confined entirely to the surface because, if this were so, the current sheet would have no thickness and the current density would be infinite, which is a physical impossibility. In fact the currents flow within a very thin surface layer whose thickness is of the order of 10^{-5} cm, although the exact value varies for different metals. We shall see that, though this thickness is so small, it plays a very important part in determining the properties of superconductors.

When a superconducting sample is in an applied magnetic field, the screening currents which circulate to cancel the flux inside must flow within this surface layer. Consequently, the flux density does not fall abruptly to zero at the boundary of the metal but dies away within the region where the screening currents are flowing. For this reason the depth within which the currents flow is called the *penetration depth*, because it is the depth to which the flux of the applied magnetic field appears to penetrate. Thus, although we speak of a superconductor as being perfectly diamagnetic, there is in fact a very slight penetration of magnetic flux, the flux density dying away at the surface as shown in

Fig. 2.6. (This is somewhat like the "skin depth" to which high frequency alternating fields penetrate in a normal conductor.) Consider the boundary of a semi-infinite slab, as shown in Fig. 2.6. If at a distance x into the metal the flux density falls to a value $B(x)$, we can define the penetration depth λ by

$$\int_0^\infty B(x)\, dx = \lambda B(0), \tag{2.2}$$

where $B(0)$ is the flux density at the surface of the metal. In other words, there would be the same amount of flux inside the superconductor if the flux density of the external field remained constant to a distance λ into the metal.

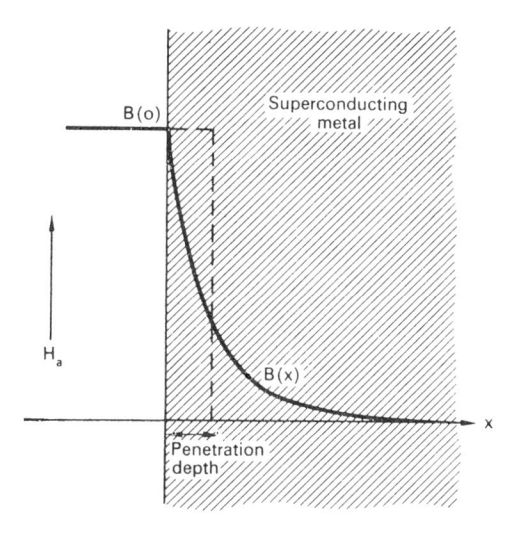

FIG. 2.6. Penetration of magnetic flux into surface of superconductor.

The London theory of superconductivity, which we shall discuss in the next chapter, predicts that in a specimen which is much thicker than the penetration depth the magnetic flux density decays exponentially as it penetrates into the metal, i.e.

$$B(x) = B(0)e^{-x/\lambda}.$$

However, in simple calculations it is often sufficient to use the approximation that the flux density $B(0)$ of the applied field remains constant to a distance λ into the metal and then abruptly falls to zero.

Because the penetration depth is so small, we do not notice the flux penetration in magnetic measurements on ordinary sized specimens,† and these appear to be perfectly diamagnetic with $\mathbf{B} = 0$. We shall, for convenience, continue to refer to reasonably large superconducting bodies as being "perfectly" diamagnetic, the very small magnetic flux at the surface being included in this term. The penetration of the magnetic flux becomes noticeable, however, if we make measurements on small samples, such as powders or thin films, whose dimensions are not much greater than the penetration depth. In these cases there is an appreciable magnetic flux density right through the metal; there is no longer perfect diamagnetism, and consequently the properties are rather different from those of the bulk superconducting metal. We shall deal fully with the special features of thin specimens in Chapter 8.

2.4.1. Variation with temperature

The penetration depth does not have a fixed value but varies with temperature, as shown in Fig. 2.7. At low temperatures it is nearly in-

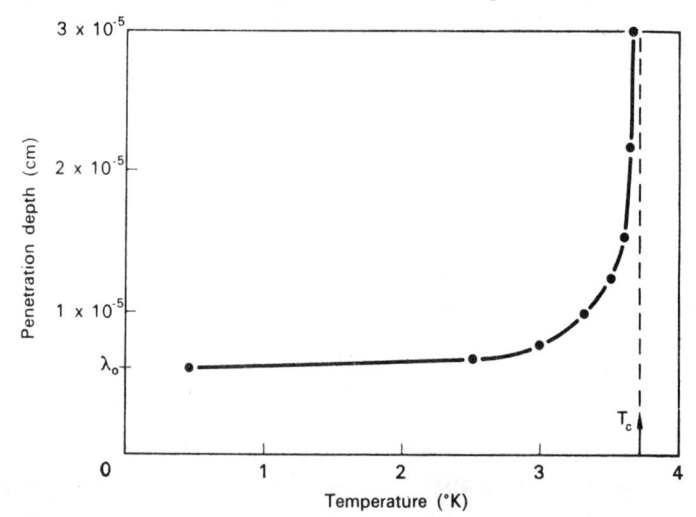

FIG. 2.7. Variation with temperature of penetration depth in tin (after Schawlow and Devlin).

dependent of temperature and has a value λ_0 characteristic of the particular metal (Table 2.1). Above about 0·8 of the transition temperature, however, the penetration depth increases rapidly and approaches infinity as the temperature approaches the transition temperature.

† The methods by which such measurements can be made are discussed in § 4.4.

The variation of penetration depth with temperature is found to fit very closely the relation

$$\lambda = \frac{\lambda_0}{(1 - t^4)}, \qquad (2.3)$$

where t is the temperature relative to the transition temperature, $t = T/T_c$.

Perfect diamagnetism does not, therefore, occur in specimens which are very close to their transition temperature. The decrease in penetration depth is, however, so rapid as the temperature falls below T_c that

TABLE 2.1. Some Values of the Penetration Depth at 0°K

	In	Al	Pb
λ_0 (cm)	$6 \cdot 4 \times 10^{-6}$	$5 \cdot 0 \times 10^{-6}$	$3 \cdot 9 \times 10^{-6}$

any *large* departure from perfect diamagnetism would be extremely difficult to detect in bulk specimens because of the difficulty of holding the temperature sufficiently constant during a measurement. For example, to observe a penetration depth of about 1 mm a specimen would, according to (2.3), have to be held at a temperature only 10^{-7} per cent below the transition temperature! Furthermore, we would require a specimen so pure and uniform that the transition to the superconducting state would take place within a temperature range which was less than 10^{-7} per cent of the transition temperature. The experimental relation (2.3) has been obtained by observing relatively small increases in the penetration depth as a superconducting metal is warmed towards its transition temperature. In one experimental arrangement[†] a rod of pure superconductor is surrounded by a closely fitting solenoid. When the temperature of the metal is lowered so that it becomes superconducting there can be no magnetic flux in the metal, except just below the surface within the penetration depth. Because the solenoid fits the rod very closely, its self-inductance is largely dependent on the magnitude of this penetration depth. A capacitor is connected across the solenoid, the combination forming the tank circuit of an oscillator of about 100 kHz frequency. When the inductance alters, because of a change in penetration depth, the oscillator frequency shifts. The frequency can be measured with great precision on an accurate frequency meter. In the

[†] A. L. Schawlow and G. E. Devlin, *Phys. Rev.* **113**, 120 (1959).

experiment of Schawlow and Devlin a frequency shift of 0·1 Hz was equivalent to a change of 4×10^{-8} cm in the penetration depth in the metal core. Such an experiment needs to be done with great care because spurious effects, not due to the change in penetration depth, may alter the resonant frequency of the solenoid and capacitor. By this method the temperature variation of the penetration depth in tin was observed, and the results are those shown in Fig. 2.7. The very rapid increase in penetration depth as the transition temperature is approached is in good accord with (2.3).

Unless we specify otherwise, when in this book we speak of the penetration depth, we mean the value λ_0 to which λ tends when the metal is at temperatures which are appreciably below its transition temperature.

The penetration depth in a superconducting metal also depends on the purity, the penetration depth increasing as the metal becomes more impure. For example, in tin containing 3% indium the penetration depth is twice that of pure tin.

CHAPTER 3

ELECTRODYNAMICS

IN THIS chapter we shall develop equations which govern the behaviour of magnetic fields and electric currents in superconductors.

3.1. Consequence of Zero Resistance

In a superconducting metal the superelectrons encounter no resistance to their motion, so, if a constant electric field **E** is maintained in the material, the electrons accelerate steadily under the action of this field:

$$m\dot{\mathbf{v}}_s = e\mathbf{E} \tag{3.1}$$

where $\mathbf{v}_s$ is the velocity of the superelectrons and m and e are their mass and charge.† If there are n_s superelectrons per unit volume moving with velocity $\mathbf{v}_s$, there is a supercurrent density

$$\mathbf{J}_s = n_s e \mathbf{v}_s.$$

Substituting this into (3.1) we see that an electric field produces a continuously increasing current with a rate of increase given by

$$\dot{\mathbf{J}}_s = \frac{n_s e^2}{m} \mathbf{E}. \tag{3.2}$$

To obtain an equation describing magnetic fields we remember that a magnetic field is related to an electric field and current by Maxwell's equations

$$\dot{\mathbf{B}} = -\operatorname{curl} \mathbf{E} \tag{3.3}$$

and
$$\operatorname{curl} \mathbf{H} = \mathbf{J} + \dot{\mathbf{D}}. \tag{3.4}$$

In this chapter we wish to develop equations which explicitly relate the fields in superconductors to the currents flowing in them, and we shall

† In this book e is used for the charge on the electron, and this symbol includes the negative sign, $e = -1 \cdot 602 \times 10^{-19}$ coulomb.

adopt the viewpoint, discussed in § 2.2.2, that the metal of a superconducting body is, like other non-ferromagnetic metals, non-magnetic, i.e. $\mu_r = 1$, so that any flux density in the metal must be due to the currents. We also take the view (see Appendix A) that while currents in the metal affect **B**, they do not affect **H**, and within the superconductor we therefore replace (3.4) by

$$\text{curl } \mathbf{B} = \mu_0(\mathbf{J}_s + \dot{\mathbf{D}}),$$

where $\mathbf{J}_s$ is the current density within the metal. Furthermore, unless the fields are varying very rapidly in time, the displacement current $\dot{\mathbf{D}}$ is negligible in comparison with $\mathbf{J}_s$. Hence inside a superconductor we can write Maxwell's equations in the form

$$\dot{\mathbf{B}} = -\text{curl } \mathbf{E}, \tag{3.3a}$$

$$\text{curl } \mathbf{B} = \mu_0 \mathbf{J}_s. \tag{3.4a}$$

Substituting (3.2) into (3.3a) gives

$$\dot{\mathbf{B}} = -\frac{m}{n_s e^2} \text{curl } \dot{\mathbf{J}}_s, \tag{3.5}$$

and we can eliminate $\mathbf{J}_s$ by means of (3.4a):

$$\dot{\mathbf{B}} = -\frac{m}{\mu_0 n_s e^2} \text{curl curl } \dot{\mathbf{B}}.$$

For brevity, let us use a single symbol α for the constant $m/\mu_0 n_s e^2$, so the last equation becomes

$$\dot{\mathbf{B}} = -\alpha \text{ curl curl } \dot{\mathbf{B}}. \tag{3.6}$$

Now

$$\text{curl curl } \dot{\mathbf{B}} = \text{grad div } \dot{\mathbf{B}} - \nabla^2 \dot{\mathbf{B}},$$

but from Maxwell's equations div $\mathbf{B} = 0$, so (3.6) becomes

$$\dot{\mathbf{B}} = \alpha \nabla^2 \dot{\mathbf{B}}$$

or
$$\nabla^2 \dot{\mathbf{B}} = \frac{1}{\alpha} \dot{\mathbf{B}}. \tag{3.7}$$

This is a differential equation which **B** must satisfy. To see what this implies, consider the plane boundary of a superconductor with a uniform magnetic field applied parallel to this boundary (Fig. 3.1). Suppose that the flux density outside the metal is $\mathbf{B}_a$, and let the direction normal to the boundary be called the x-direction. Because the applied field is uniform, $\mathbf{B}$ will have the same direction everywhere, so we may regard (3.7) as a scalar equation; also there will be no gradients of the field parallel to the boundary, so in this case (3.7) reduces to

$$\frac{\partial^2 \dot{B}}{\partial x^2} = \frac{1}{a}\dot{B}. \qquad (3.8)$$

The solution of this is†

$$\dot{B}(x) = \dot{B}_a \exp\left(\frac{-x}{\sqrt{a}}\right), \qquad (3.9)$$

where $\dot{B}(x)$ is the flux density at a distance x inside the metal and $\dot{B}_a$ is the value of $\dot{B}$ outside the metal. This means that $\dot{B}$ dies away exponent-

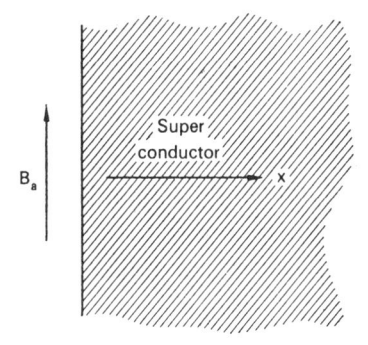

FIG. 3.1. Magnetic field applied parallel to boundary of superconductor.

ially as we penetrate into the superconductor. In other words, changes in flux density do not penetrate far below the surface, so at a sufficient distance inside the metal the flux density has a constant value which does not change with time, irrespective of what is happening to the applied field.

3.2. The London Theory

We have deduced the above behaviour by applying the usual laws of electrodynamics to a conductor with zero resistance; however, though (3.9) completely describes the magnetic properties of a perfect conduct-

† There is an alternative solution, $B_a \exp(+x/\sqrt{a})$, but this approaches infinity as x increases.

or, it does not adequately describe the behaviour of a superconductor. The Meissner effect shows that inside a superconductor the flux density is not only constant but the value of this constant is always zero; so not only $\dot{B}$ but B itself must die away rapidly below the surface. F. and H. London[†] suggested that the magnetic behaviour of a superconducting metal might be correctly described if (3.7) applied not only to $\dot{\mathbf{B}}$ but to $\mathbf{B}$ itself,

$$\nabla^2 \mathbf{B} = \frac{1}{a}\mathbf{B}. \qquad (3.10)$$

If this were so, the magnetic flux density B would die away within the metal in a manner similar to the behaviour of $\dot{B}$, as described by (3.9),

i.e. $\qquad B(x) = B_a \exp(-x/\sqrt{a}).$

Examination of the argument by which we derived (3.7) shows that we could have derived (3.10) if everywhere we had replaced $\dot{\mathbf{B}}$ by $\mathbf{B}$. If we retrace the argument, we arrive at (3.5) which would now have the more restrictive form

$$\mathbf{B} = \frac{-m}{n_s e^2} \operatorname{curl} \mathbf{J}_s. \qquad (3.11)$$

This equation and (3.2), namely

$$\dot{\mathbf{J}}_s = \frac{n_s e^2}{m} \mathbf{E}, \qquad (3.2)$$

which together describe the electrodynamics of the supercurrent, are known as the *London equations*. Equation (3.2) describes the resistanceless property of a superconductor, there being no electric field in the metal unless the current is changing; and (3.11) describes the diamagnetism.

Note that these equations are not deduced from fundamental properties and do not "explain" the occurrence of superconductivity. The London equations are restrictions on the ordinary equations of electromagnetism, and are introduced so that the behaviour deduced from these laws agrees with that observed experimentally.

The London equations lead us to replace (3.7) by (3.10). Let us now use this latter equation to determine the distribution of magnetic flux inside a superconductor when it is in a uniform magnetic field of flux den-

---

[†] F. London and H. London, *Proc. Roy. Soc.* (*London*) **A155**, 71 (1935).

sity B_a applied parallel to its surface (Fig. 3.2). In this case we can use the one-dimensional form of (3.10),

$$\frac{\partial^2 B(x)}{\partial x^2} = \frac{1}{\alpha} B(x),$$

where $B(x)$ is the flux density at distance x inside the metal. This equation has the solution

$$B(x) = B_a \exp\left(\frac{-x}{\sqrt{\alpha}}\right), \qquad (3.12)$$

where B_a is the flux density of the applied field at the surface. Equation (3.12) shows that the flux density dies away exponentially inside a superconductor, falling to $1/e$ of its value at the surface at a distance $x = \sqrt{\alpha}$. This distance is called the *London penetration depth*, λ_L. Now $\alpha = m/\mu_0 n_s e^2$, so the London penetration depth is given by

$$\lambda_L = \sqrt{(m/\mu_0 n_s e^2)}. \qquad (3.13)$$

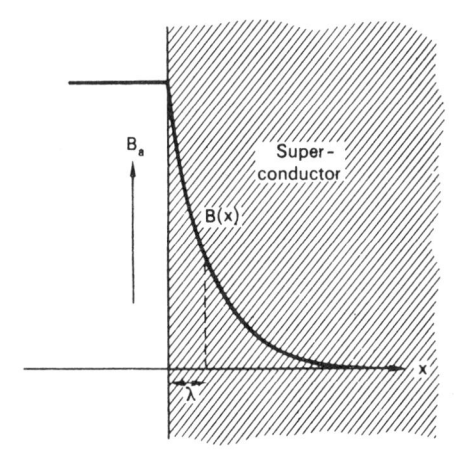

FIG. 3.2. Variation of flux density at boundary of a superconductor.

If we substitute for m and e the usual values of the electron's mass and charge, and take n_s to be about 4×10^{28} m^{-3} (i.e. the usual concentration in metals; about one conduction electron per atom), the London penetration depth turns out to be about 10^{-6} cm.

We can now write (3.12) in the form

$$B(x) = B_a \exp(-x/\lambda_L). \qquad (3.14)$$

The London equations predict, therefore, a very rapid exponential decay of the flux density at the surface of a superconductor. In Chapter 2 we defined, by means of (2.2), a general penetration depth which was independent of the particular form of the decrease of the flux density (i.e. whether exponential or otherwise). Substitution of (3.14) into (2.2) shows that the London penetration depth λ_L, defined by (3.13) and leading to an exponential penetration law of the form (3.14), satisfies this general definition of penetration depth.

In the previous chapter it was pointed out that any current in a superconductor must flow near the surface. For a uniform magnetic field applied parallel to the surface (Fig. 3.1) in the z-direction, (3.4a) reduces to $-\frac{\partial B}{\partial x} = \mu_0 \mathcal{J}_y$. From (3.14) $\partial B/\partial x$ equals $-\frac{B_a}{\lambda_L} \exp(-x/\lambda_L)$, so we have

$$\mathcal{J}_y = \frac{B_a}{\mu_0 \lambda_L} \exp(-x/\lambda_L)$$

which we may write as

$$\mathcal{J}_y = \mathcal{J}_a \exp(-x/\lambda_L).$$

We see, therefore, that any current flows close to the surface within the penetration depth.

We have mentioned that according to the two-fluid model of superconductivity (§ 1.4), the concentration of superelectrons n_s decreases as the temperature is raised, falling to zero at the transition temperature. The London equations give a penetration depth which is inversely proportional to $n_s^{\frac{1}{2}}$ (3.13) and so predict that the penetration depth should increase with increasing temperature, rising to infinity as the temperature approaches the transition temperature. As we saw in § 2.4.1, this is what is observed experimentally.

We can now write the London equations (3.11) and (3.2) as

$$\operatorname{curl} \mathbf{J}_s = -\frac{1}{\mu_0 \lambda_L^2} \mathbf{B}$$

$$\dot{\mathbf{J}}_s = \frac{1}{\mu_0 \lambda_L^2} \mathbf{E}.$$

It is important to realize that the London equations do not replace Maxwell's equations, which, of course, still apply to all currents and the fields they produce. The London equations are additional conditions obeyed by the supercurrents.

In the most general case the total current **J** is the sum of a normal current and a supercurrent:

$$\mathbf{J} = \mathbf{J}_n + \mathbf{J}_s.$$

The normal current need only obey Maxwell's equations and Ohm's law,

$$\mathbf{J}_n = \sigma' \mathbf{E},$$

where σ' is a conductivity associated with the normal electrons. We can now bring together the special equations which apply to a superconducting metal:

$$\mathbf{J} = \mathbf{J}_n + \mathbf{J}_s \qquad (3.15)$$

$$\mathbf{J}_n = \sigma' \mathbf{E} \qquad (3.16)$$

$$\operatorname{curl} \mathbf{J}_s = -\frac{1}{\mu_0 \lambda_L^2} \mathbf{B} \qquad (3.17)$$

$$\dot{\mathbf{J}}_s = \frac{1}{\mu_0 \lambda_L^2} \mathbf{E}. \qquad (3.18)$$

From these equations we can in principle work out the distribution of fields and currents in superconducting bodies under various conditions. In the steady state, when fields and currents are not changing with time, the only current is the supercurrent, i.e. $\mathbf{J}_n = 0$, and we need only employ the London equations (3.17) and (3.18). These lead to

$$\nabla^2 \mathbf{B} = \frac{1}{\lambda_L^2} \mathbf{B}. \qquad (3.19)$$

It should be pointed out that the form of (3.10) introduced to describe the distribution of magnetic flux density inside a superconductor was only a guess, based on the known properties of superconductors, and we must not expect the London equations inferred from it to be exactly correct. In fact, these equations are only approximations, though for many purposes they are sufficiently accurate. For example, the London equations predict a small penetration depth; a small penetration depth is indeed observed experimentally, but the value is greater than the London prediction by a factor of two or more. A discussion of the limitations of the London theory and a description of the more refined Ginzburg–Landau theory are given in Chapter 8.

3.2.1. An application of the London theory

We can in principle use (3.19) to find the distribution of flux density within any superconducting body by applying to the solution of this equation boundary conditions imposed by the shape of the body and the form of the applied magnetic field. As an example, we now calculate the distribution of flux density inside a superconducting slab, of finite thickness, with plane parallel faces, when it is in a uniform magnetic field applied parallel to these faces (Fig. 3.3). This is a configuration we shall

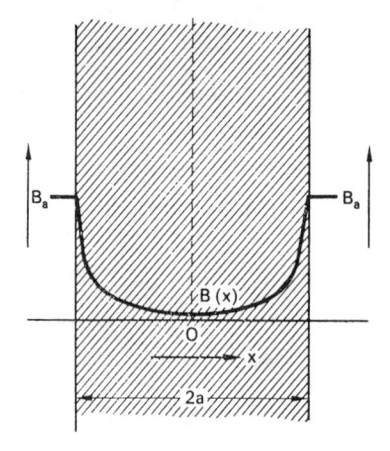

FIG. 3.3. Superconducting slab in parallel applied magnetic field.

need to make use of in a later chapter (Chapter 8). Let the thickness of the slab be $2a$ and let the x-coordinate be normal to the faces, with the origin at the mid-plane. Because the applied field is uniform,

$$\frac{\partial B}{\partial y} = 0 = \frac{\partial B}{\partial z}$$

and we can write (3.19) as

$$\frac{\partial^2 B(x)}{\partial x^2} = \frac{1}{\lambda_L^2} B(x).$$

The solution of this equation is

$$B(x) = B_1 \, e^{+x/\lambda_L} + B_2 \, e^{-x/\lambda_L}. \tag{3.20}$$

The values of the constants B_1 and B_2 are obtained by noting that, from symmetry, the variation of B in both the $+x$ and $-x$ directions must be the same. Therefore $B_1 = B_2$. Further, when $x = \pm a$, the flux density equals that of the applied field, $B(\pm a) = B_a$. So (3.20) becomes

$$B(x) = \frac{B_a}{\cosh(a/\lambda_L)} \cosh(x/\lambda_L). \qquad (3.21)$$

The flux density therefore varies as $\cosh(x/\lambda_L)$. The width of most specimens will be much greater than the penetration depth λ_L and then the flux density will be very small except close to the faces where $|x| \to |a|$.

CHAPTER 4

THE CRITICAL MAGNETIC FIELD

WE SHALL see in Chapter 9 that, if a metal is to remain superconducting, the net momentum of the superelectrons must not exceed a certain value. For this reason there is a limit to the density of resistanceless current† that can be carried by any region in the metal. Let us call this the *critical current density* J_c of the metal. This critical current density applies both to a current passed along the specimen from an external source and to screening currents which shield the specimen from an applied magnetic field. We now show that, as a result of this critical current density, a superconducting metal will be driven normal if a sufficiently strong magnetic field is applied to it.

As we have seen, the perfect diamagnetism of a superconductor arises because, in an applied magnetic field, resistanceless surface currents circulate so as to cancel the flux density inside. If the strength of the applied magnetic field is increased, the shielding currents must also increase in order to maintain perfect diamagnetism. If the applied magnetic field is increased sufficiently, the critical current density will be reached by the shielding currents and the metal will lose its superconductivity. The shielding currents then cease and the flux due to the applied magnetic field is no longer cancelled within the metal.‡ There is therefore a limit to the strength of magnetic field which can be applied to a superconductor if it is to remain superconducting. This destruction of superconductivity by a sufficiently strong magnetic field is one of the most important properties of a superconductor.

At any point in a superconductor there is a definite relationship

---

† Strictly speaking it is not the current but the metal which is resistanceless. However, the expression "resistanceless current" is often used and we shall make use of this convenient phrase to avoid circumlocution.

‡ The transition from the perfectly diamagnetic superconducting state to the non-magnetic normal state is a *reversible* transition. The screening currents do not die away with dissipation of energy and do not generate heat in the material. In fact, as we shall see from thermodynamic arguments in § 5.2.2, if a thermally isolated specimen is driven normal by a magnetic field its temperature *falls*.

between the supercurrent density $\mathcal{J}_s$ and the magnetic flux density, though the values of these are only appreciable within the penetration depth. This relationship can be obtained from the London equations (Chapter 3). The critical current density will therefore be reached at the surface when the flux density of the applied magnetic field (i.e. the flux density at the surface) is increased to a certain value. We may call this the "critical flux density" B_c. However, the flux density *outside* the metal always equals $\mu_0 H$, where H is the magnetic field strength, so we may equally well refer to a *critical magnetic field strength*, $H_c = B_c/\mu_0$. Notwithstanding our view that B rather than H is the basic magnetic quantity (see Appendix A), it is usual in the literature to refer to the critical magnetic field strength H_c rather than the critical flux density B_c. In this book we shall follow this convention, and refer to the critical magnetic field strength H_c remembering that this is always related to the critical flux density by $H_c = B_c/\mu_0$.

4.1. Free Energy of a Superconductor

We saw in § 2.2.1 that the state of magnetization of a superconductor depends only on the *values* of the applied magnetic field and temperature and not on the way these external conditions were arrived at. This implies that, whether or not there is an applied magnetic field, the transition from the superconducting to the normal state is reversible, in the thermodynamic sense. We may therefore apply thermodynamic arguments to a superconductor, using the temperature and magnetic field strength as thermodynamic variables.

It is possible to deduce something about the critical magnetic field by considering what effect the application of a magnetic field has on the free energy of a superconducting specimen. We are interested in the free energy because in any system the stable state is that with the lowest free energy. In considering the critical magnetic field of a superconductor we are interested in the *Gibbs* free energy, because we want to compare the difference in the magnetic contribution to the free energy of two phases, superconducting and normal, when they are in the same applied magnetic field (i.e. with the *intensive* thermodynamic variable constant).

Consider a specimen of superconductor in the form of a long rod. (At this stage we consider a long rod so as to reduce to a negligible degree special demagnetizing effects which are due to the ends of the specimen. These effects are discussed in § 6.2.) When the specimen is cooled below its transition temperature it becomes superconducting. Therefore, below the transition temperature the free energy of the superconducting state must be less than that of the normal state, otherwise the metal would remain normal. Suppose that at a temperature T, and in the absence of an applied magnetic field ($H_a = 0$), the Gibbs free energy per unit volume of the superconducting state is $g_s(T, 0)$ and that of the normal state is $g_n(T, 0)$ (Fig. 4.1). Now let a magnetic field of strength H_a be applied parallel

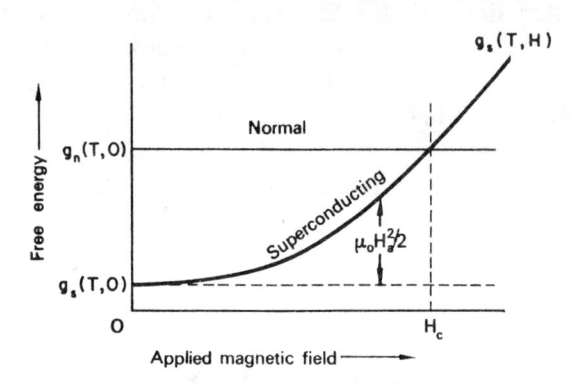

FIG. 4.1. Effect of applied magnetic field on Gibbs free energy of normal and superconducting states.

to the length of the rod. Any substance, which in an applied field H_a acquires a magnetization I, changes its free energy per unit volume by an amount†

$$\Delta g(H_a) = -\mu_0 \int_0^{H_a} I \, dH_a. \tag{4.1}$$

So in the case of the field producing a positive magnetization, i.e. the magnetization in the same direction as the magnetic field, the free energy is lowered. [Note that (4.1) implies that, when a substance is magnetized by an applied magnetic field, its free energy changes by an amount proportional to the area under its magnetization curve (I v. H_a). This is a useful general result, which we shall employ several times.] In the case of a superconducting specimen the application of a magnetic field produces a *negative* magnetization which, if penetration of the field is neglected, exactly cancels the flux due to the applied field, so that $I = -H$. The free energy per unit volume is therefore *increased* to

$$g_s(T, H) = g_s(T, 0) + \mu_0 \int_0^{H_a} |I| \, dH_a.$$

† See Appendix B.

In fact $|I| = H$, so the magnetic field raises the free energy density to

$$g_s(T, H) = g_s(T, 0) + \mu_0 \frac{H_a^2}{2}. \tag{4.2}$$

So, when we apply a magnetic field to a superconductor, its free energy increases to this value due to the magnetization (Fig. 4.1). The *normal* state, however, is virtually non-magnetic and acquires negligible magnetization in an applied magnetic field. Consequently the application of a magnetic field does not change the free energy of the normal state though it raises that of the superconducting state. If the field strength is increased enough, the free energy of the superconducting state will be raised above that of the normal state, and, in this case, the metal will not remain superconducting but will become normal. This occurs when $g_s(T, H) > g_n(T, 0)$, which with (4.2) gives

$$\mu_0 \frac{H_a^2}{2} > [g_n(T, 0) - g_s(T, 0)].$$

There is therefore a maximum magnetic field strength that can be applied to a superconductor if it is to remain in the superconducting state. This critical magnetic field strength is given by

$$H_c(T) = \left\{ \frac{2}{\mu_0} [g_n(T, 0) - g_s(T, 0)] \right\}^{\frac{1}{2}}. \tag{4.3}$$

This critical magnetic field which we have derived by a thermodynamic argument is the same as the critical field which we discussed on p. 40 in terms of a critical current density.

The critical magnetic field strength can be measured quite simply by applying a magnetic field parallel to a wire of superconductor and observing the strength at which resistance appears.

4.2. Variation of Critical Field with Temperature

If the critical magnetic field of a superconductor is measured, its value is found to depend on the temperature (Fig. 4.2), falling from some value H_0, at very low temperatures, to zero at the superconducting transition temperature T_c. A diagram such as Fig. 4.2 can be called the *phase diagram* of a superconductor. The metal will be superconducting for any combination of temperature and applied field which gives a point, such as P, lying within the shaded region. As the arrows indicate, the metal can be driven into the normal state by increasing either the temperature or the applied field, or both.

44   INTRODUCTION TO SUPERCONDUCTIVITY

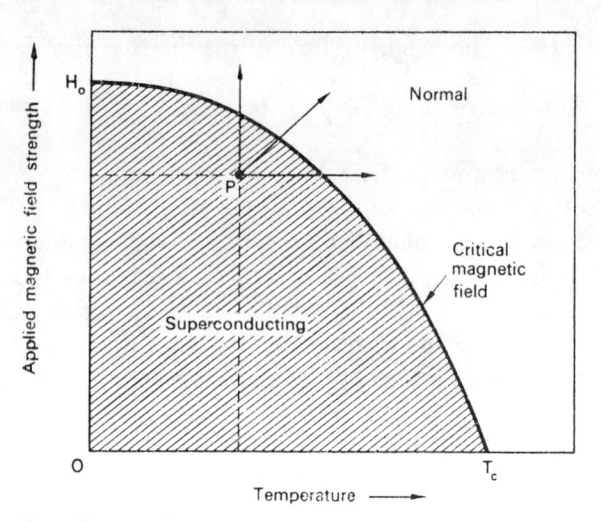

FIG. 4.2. Phase diagram of a superconductor, showing variation with temperature of the critical magnetic field.

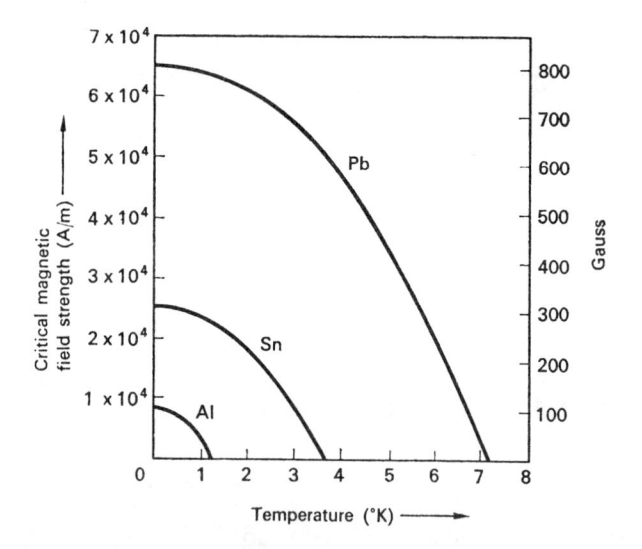

FIG. 4.3. Critical fields of some superconductors.

The value of H_0 is different for each superconducting metal; metals with low transition temperatures have low critical fields at absolute zero. So each superconductor has a different phase diagram (Fig. 4.3). From experiment it has been found that the critical fields fall off almost as the square of the temperature, so the critical field curves are closely approximated by parabolas of the form

$$H_c = H_0 [1 - (T/T_c)^2], \tag{4.4}$$

where H_0 is the critical field at absolute zero and T_c is the transition temperature. Each superconductor can be characterized by its particular values of H_0 and T_c, and, knowing these, we can use (4.4) to find the critical field at any temperature. Table 1.1 lists values of H_0 and T_c for the superconducting elements.

We may remark here that there is no fundamental significance in the relation between critical field and temperature being a parabola. It has merely been found experimentally that the variation can be conveniently described to within a few per cent by a parabolic curve. The experimental curves are not in fact exactly parabolas, and to describe them accurately one would need a polynomial expression. For most calculations, however, it is sufficient to use the parabola given by (4.4).

The Cryotron

The existence of a critical magnetic field has been made use of in a controlled switch called a *cryotron* (Fig. 4.4). The current $\mathscr{I}$ which is to be controlled flows along a straight wire of tantalum called the "gate".

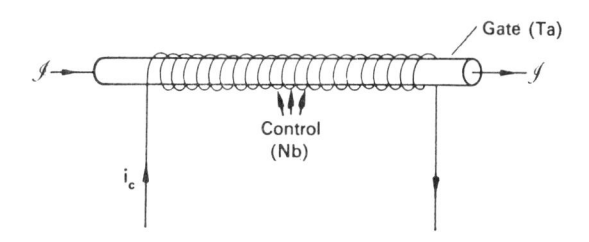

FIG. 4.4. Cryotron of tantalum and niobium.

Around this, but insulated from it, is a niobium wire wound in a long single-layer coil called the "control". When cooled to 4·2°K by immersion in liquid helium both the tantalum gate and the niobium control are superconducting, and the gate offers no resistance to the passage of the current $\mathscr{I}$. If, however, a current i_c is passed through the control coil, it generates a magnetic field along the gate, and, if the control current is large enough, the gate is driven normal by the magnetic field, and the appearance of resistance reduces the current $\mathscr{I}$. The control coil, however, remains resistanceless because the critical field of niobium is considerably higher than that of tantalum. Hence the current $\mathscr{I}$ through the gate can be controlled by a smaller current in the control, and the device is analogous to a relay. Small cryotrons were first developed as fast acting switches for possible use in digital computers. Large cryotrons can be used to control the currents in superconducting magnet circuits (see § 1.3).

4.3. Magnetization of Superconductors

We now discuss how the magnetization of a superconducting specimen varies as a magnetic field of increasing strength is applied. Let us again consider a rod of superconductor and imagine a magnetic field H_a to be applied parallel to its length. Figure 4.5a shows how the flux density B inside the specimen varies as we increase the strength of the

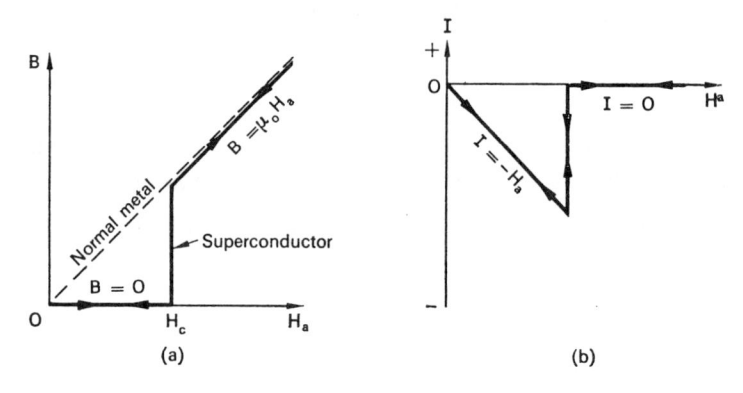

FIG. 4.5. Magnetic behaviour of a superconductor.

applied field. Normal metals (excluding the special ferromagnetic metals, such as iron) are virtually non-magnetic and so the flux density B inside them is proportional to the strength of the applied field, $B = \mu_0 H_a$, as shown by the dotted line in Fig. 4.5a. A superconductor, however, is perfectly diamagnetic if we neglect the penetration depth, and as the applied magnetic field strength is increased, the flux density within the specimen remains at zero. But when the applied field strength reaches the critical value H_c, the superconductor is driven into the normal state, and the flux due to the applied field is no longer cancelled inside. At all higher applied field strengths the superconductor behaves just like a normal metal. For a pure specimen this behaviour is reversible; if the applied magnetic field is decreased from a high value, the specimen goes back into the superconducting state at the value H_c and below this there is no net flux inside.

THE CRITICAL MAGNETIC FIELD

We can describe the magnetic behaviour of a superconductor in another way. We have seen that, when a metal is in the superconducting state, there is no magnetic flux inside, because surface currents circulate to give the specimen a magnetization I exactly equal and opposite to the applied field, so that $I = -H_a$. Figure 4.5b shows how the magnetization of a superconductor varies with the strength of the applied magnetic field. When the applied field strength reaches H_c the superconductor becomes normal and the negative magnetization disappears. At higher applied fields the superconductor has, like any normal metal, virtually no magnetization. Figure 4.5 is, of course, just two equivalent ways of presenting the same information. We need to be familiar with the form of both curves, because sometimes it is convenient to consider the internal flux density, whereas on other occasions it is more convenient to consider magnetization.

4.3.1. "Non-ideal" specimens

The magnetic properties we have been considering so far in this chapter are those that would be shown by "ideal" specimens, i.e. those containing no impurities or crystalline faults. Any real specimen is, however, not perfect, and its behaviour will depart to some extent from the behaviour we have just described. Nevertheless, it is possible with great care to produce specimens so nearly perfect that they have properties very closely approximating to the ideal. However, the greater the degree of imperfection the greater will be the departure from ideal behaviour.

An ideal specimen has a sharply defined critical field strength and its magnetization curve is completely reversible. Figure 4.6 illustrates the

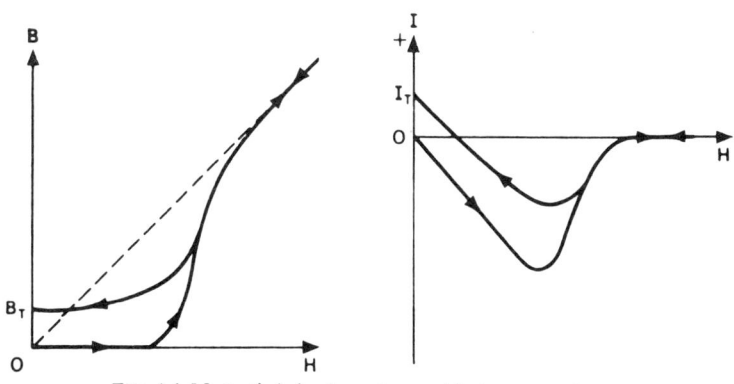

FIG. 4.6. Magnetic behaviour of a non-ideal superconductor.

magnetic behaviour of an imperfect sample. It can be seen that there is no longer a sharply defined critical magnetic field, the transition from the superconducting to normal state being "smeared out" over a range of applied field strengths. Furthermore, the magnetization is not reversible; in decreasing fields the curves follow different paths from those traced in the original increasing field. We call this *hysteresis*. Finally, when the applied field has been reduced to zero, there may remain some positive magnetization of the sample, giving rise to a residual flux density B_T and magnetization I_T. We say the sample has *trapped flux*. In this condition the superconductor is like a permanent magnet.

We see, therefore, that a non-ideal specimen may show:

1. Ill-defined critical magnetic field.
2. Magnetic hysteresis.
3. Trapped flux.

These three departures from ideal behaviour do not necessarily all occur together. For example, a specimen may not have a sharp critical field and may show hysteresis but still not trap any flux. Defects involving large numbers of atoms, such as particles of another substance or the chains of displaced atoms known as dislocations, tend to give rise to hysteresis and trapped flux, whereas impurity atoms and unevenness of composition reduce the sharpness of the critical field. The reasons why different kinds of impurity and imperfection produce the various departures from ideal behaviour are complicated and not yet fully understood, and so we shall not discuss them in detail here. However, these effects are of considerable practical importance and we shall return to them again in Chapter 12.

4.4. Measurement of Magnetic Properties

The techniques which can be used to measure the magnetic characteristics of superconductors do not differ in principle from those used in measurements on ordinary magnetic materials, such as a ferromagnetics; but they must, of course, be suitable for use at very low temperatures. The methods fall into two classes: those which measure the flux density B in the sample and those which measure the sample's magnetization I (Fig. 4.5). Though either of these gives the full information on the magnetic properties of the sample, it may be convenient to choose one or other method of measurement, depending on circumstance. Many different types of apparatus have been used, of various

4.4.1. Measurement of flux density

This is a very simple measurement which does not require any moving parts in the low temperature region. The principle is to measure the magnetic flux which appears in a specimen when a magnetic field is applied. On the specimen X (Fig. 4.7) is wound a "pick-up" coil C consisting of a few hundred turns of fine wire. The ends of this coil are connected to a ballistic galvanometer G outside the low temperature apparatus. A magnetic field H can be applied parallel to the axis of the specimen by means of the solenoid S, and when the magnetic field is suddenly applied by closing the switch, the ballistic galvanometer will be deflected by an amount proportional to the flux threading the pick-up coil C, i.e. the deflection is proportional to the flux density B. Hence by successively applying stronger magnetic fields we can follow the varia-

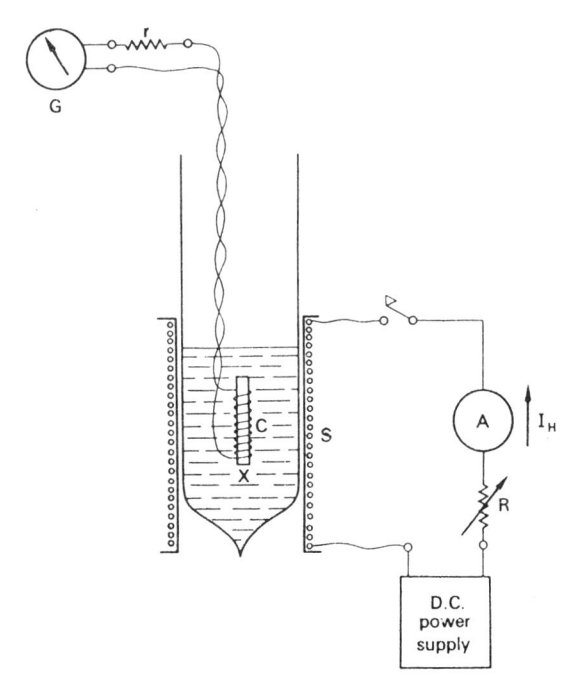

FIG. 4.7. Measurement of magnetic flux density in a superconductor.

tion of B with field strength H. Figure 4.8 shows a B versus H characteristic obtained in this way. The critical field H_c at which the specimen ceases to be perfectly diamagnetic can clearly be seen.

In the case of a non-ideal specimen some flux will be trapped and so, when the field is switched off, there will be a smaller reverse deflection of

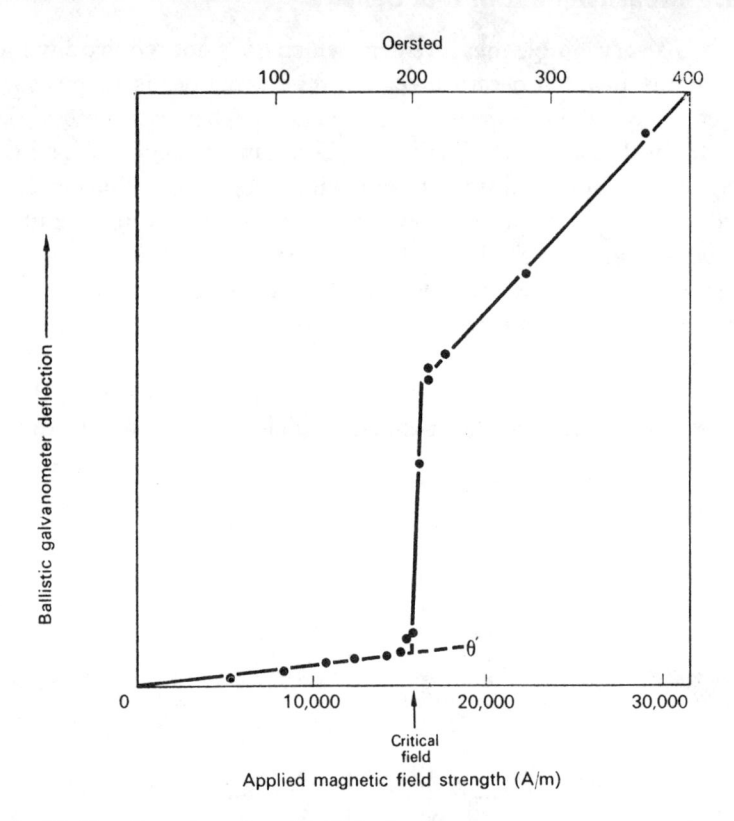

FIG. 4.8. Experimental results of ballistic flux measurement on a superconductor (tantalum at 3·7°K). The small "background" deflection θ' is due to the flux density produced by the applied field in the actual wire of the pick-up coil (after Rose-Innes).

the galvanometer. It is therefore necessary to note the galvanometer deflection both when switching on and when switching off the applied magnetic field in order to observe how much flux is trapped.

4.4.2. Measurement of magnetization

In this method the specimen is moved into and out of a pick-up coil which is connected to a ballistic galvanometer. The throw of the galvanometer will be proportional to the flux carried by the sample and this will be proportional to its magnetization. The method is illustrated in Fig. 4.9. The specimen X is mounted on the end of a sliding rod R so that it can be moved from the upper pick-up coil A to the lower one B. A and B are identical coils, except that they are wound in opposite direc-

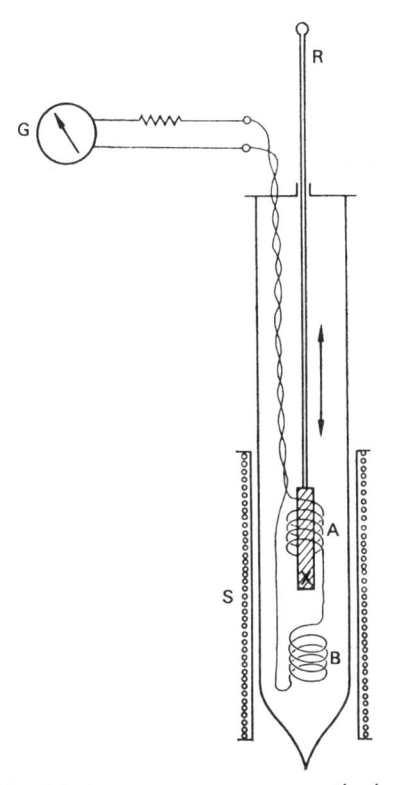

FIG. 4.9. Apparatus to measure magnetization.

tions. To measure the magnetization of the sample a steady magnetic field of the required magnitude is applied by means of the solenoid S and the specimen is then moved quickly from coil A into coil B. In the case of a diamagnetic sample, the flux threading coil A will therefore increase and the flux through B will decrease. Since the two coils are connected in series opposition the e.m.f.s. induced in them will add together and the ballistic galvanometer will swing by an amount proportional to the magnetization of the sample. The measurement is repeated at gradually increasing applied magnetic field strengths, and so we may construct a graph of sample magnetization as a function of applied field strength.

Because A and B are wound in opposite directions any e.m.f.s. due to unintended fluctuations of the applied field strength should cancel and not deflect the galvanometer.

A feature of measurements on superconductors is that they are self-calibrating. At applied field strengths below the critical field strength, the slope of the I versus H curve always equals -1.

4.4.3. Integrating method

In this method, illustrated in Fig. 4.10, we again have two identical pick-up coils A and B connected in series opposition, but the specimen is now permanently located in one of them. The current to the solenoid S is

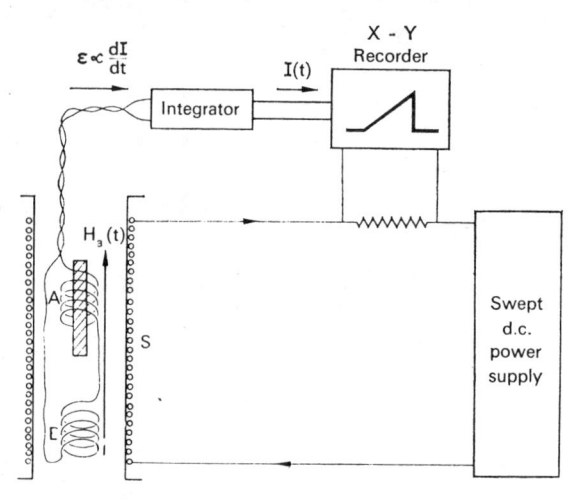

FIG. 4.10. Integration method of measuring magnetization.

gradually and smoothly increased so that a continuously increasing magnetic field is applied to the pick-up coils. An e.m.f. will be developed across each coil, proportional to the rate of change of the magnetic flux threading it, but, since one coil contains the sample and the other does not, and since they are in series opposition, the net e.m.f. $\mathscr{E}$ across the two coils will be proportional to the rate of change of magnetization of the sample:

$$\mathscr{E} = \mathscr{E}_A - \mathscr{E}_B \propto \frac{d}{dt}\mu_0(H+I) - \frac{d}{dt}\mu_0 H$$

or

$$\mathscr{E} \propto \frac{dI}{dt}.$$

This e.m.f. is fed into an electronic integrator, a circuit whose output voltage is related to its input voltage by $V_{\text{out}} \propto \int V_{\text{in}} \, dt$. The output of the integrator is therefore

$$V_{\text{out}} \propto \int_0^t \frac{dI}{dt} \, dt = I,$$

so at any time the output voltage of the integrator is proportional to the magnetization of the sample.

This method of measuring magnetization has the advantage that no movement of the specimen is required, but the electronic equipment is more complicated than the apparatus described in the previous two sections.

CHAPTER 5

THERMODYNAMICS OF THE TRANSITION

IN PREVIOUS chapters we have from time to time made use of thermodynamic arguments to derive some of the properties of superconductors. In particular, we have been able to learn much about critical magnetic fields by considering how the free energy of a superconductor is altered by the application of a magnetic field. In this chapter we discuss a few further thermodynamic aspects not treated elsewhere in this book.

5.1. Entropy of the Superconducting State

We saw in Chapter 4 that though the free energy density g_n of a metal in the normal state is independent of the strength H_a of any applied magnetic field, the application of a magnetic field raises the free energy density g_s of the metal in the superconducting state by an amount $\frac{1}{2}\mu_0 H_a^2$. The critical field H_c is that field strength which would be required to raise the free energy of the superconducting state above that of the normal state. We have, therefore, in an applied magnetic field of strength H_a a difference in free energy between the normal and superconducting states,

$$g_n - g_s(H_a) = \tfrac{1}{2}\mu_0(H_c^2 - H_a^2). \tag{5.1}$$

As shown in Appendix B, the free energy of a magnetic body can be written

$$G = U - TS + pV - \mu_0 H_a M,$$

where U is the internal energy, S the entropy, p the pressure, V the volume, H_a the applied magnetic field and M the magnetic moment. If the pressure and applied field strength are kept constant but the temperature is varied by an amount dT there will be a change of free energy,

$$dG = dU - TdS - SdT + pdV - \mu_0 H_a dM.$$

But, by the first law of thermodynamics,

$$dU = TdS - pdV + \mu_0 H_a dM$$

so $\quad dG = -SdT \quad$ and $\quad S = -\left(\dfrac{\partial G}{\partial T}\right)_{p, H_a}$

The entropy per unit volume is given by

$$s = -\left(\dfrac{\partial g}{\partial T}\right)_{p, H_a}.$$

Substituting (5.1) into this we obtain for a superconductor,† since H_a does not depend on T,

$$s_n - s_s = -\mu_0 H_c \dfrac{dH_c}{dT}. \tag{5.2}$$

Now the critical magnetic field always decreases with increase of temperature, so dH_c/dT is always negative and the right-hand side of this equation must be positive. Hence, by simple thermodynamic reasoning applied to the known variation of critical field with temperature, we have been able to deduce that the entropy of the superconducting state is less than that of the normal state, i.e. that the superconducting state has a higher degree of order than the normal state. This is in agreement with the BCS microscopic description of superconductivity (Chapter 9) according to which the electrons in a superconductor "condense" into a highly correlated system of electron pairs.

The critical field H_c falls to zero as the temperature is raised towards T_c, therefore, according to (5.2), the entropy difference between the normal and superconducting states vanishes at this temperature. Furthermore, by the third law of thermodynamics, s'_n must also equal s_s at $T = 0$. An example of the temperature variation of the entropies is shown in Fig. 5.1.

From the fact that the entropies of the superconducting and normal states must be the same at $T = 0$ we can deduce from (5.2) that, since the critical field H_c is not zero, dH_c/dT must be zero at 0°K. This is in accordance with the experimental observation that, for all superconduc-

---

† H_a does not appear in (5.2) and, as we have assumed the normal state to be non-magnetic (i.e. properties independent of any applied field), this equation implies that the entropy of the superconducting state is independent of any applied magnetic field. This is only strictly true if we ignore the flux within the penetration depth. Equation (5.1), from which (5.2) is derived, did not take account of any flux penetration into the superconducting state. Equation (5.2) applies to bulk samples, i.e. those with dimensions greater than the penetration depth.

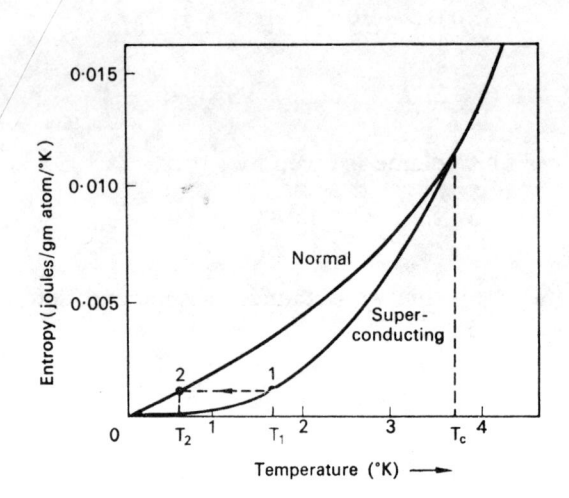

FIG. 5.1. Entropy of normal and superconducting tin (based on Keesom and van Laer). T_1 and T_2 refer to the adiabatic magnetization process described in § 5.2.2.

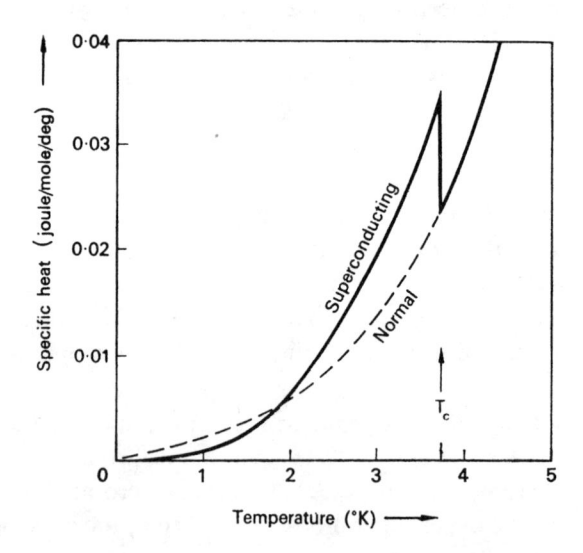

FIG. 5.2. Specific heat of tin in normal and superconducting states.

tors, the slope of the H_c versus T curve (Fig. 4.2) appears to become zero as the temperature approaches 0°K.

5.2. Specific Heat and Latent Heat

Much of the understanding of superconductors has been derived from measurements of their specific heat. The solid curve in Fig. 5.2 shows how the specific heat of a typical type-I superconductor varies with temperature in the absence of any applied magnetic field. We can draw several conclusions from the form of this curve, particularly if we compare it with the specific heat curve of the same metal in the normal state. The curve for the normal state can be obtained by making measurements in an applied magnetic field strong enough to drive the superconductor normal.

5.2.1. First-order and second-order transitions

The form of the specific heat curves can be predicted from thermodynamic arguments.

Since, at the transition temperature, $s_n = s_s$ and we have shown that $s = -(\partial g/\partial T)_{p,H_a}$, we have, for the superconducting-normal transition at T_c,

$$\left(\frac{\partial g}{\partial T}\right)_n = \left(\frac{\partial g}{\partial T}\right)_s.$$

A phase transition which satisfies this condition (i.e. not only is g continuous but $\partial g/\partial T$ is also continuous) is known as a *second-order phase transition*. A second-order transition has two important characteristics: at the transition there is no latent heat, and there is a jump in the specific heat.[†] The first characteristic follows immediately from the fact that $dQ = Tds$ and we have seen that at the transition temperature $s_n = s_s$. Hence when the transition occurs there is no change in entropy and therefore no latent heat. The second condition follows from the fact that the specific heat of a material is given by

$$C = vT\frac{\partial s}{\partial T}, \qquad (5.3)$$

[†] For a discussion of second-order phase transitions, see Pippard, *Classical Thermodynamics*, Cambridge University Press, 1961, Ch. 9.

where v is the volume per unit mass, so the difference in the specific heats of the superconducting and normal states can be obtained from (5.2):

$$C_s - C_n = vT\mu_0 H_c \frac{d^2 H_c}{dT^2} + vT\mu_0 \left(\frac{dH_c}{dT}\right)^2. \qquad (5.4)$$

In particular, at the transition temperature, $H_c = 0$ and so we have for the transition in the absence of an applied magnetic field

$$(C_s - C_n)_{T_c} = vT_c\mu_0 \left(\frac{dH_c}{dT}\right)^2_{T_c}. \qquad (5.4a)$$

This is known as Rutgers' formula, and it predicts the value of the discontinuity in the specific heat of a superconductor at the transition temperature. It should be emphasized that, though (5.4a) contains a term depending on H_c, it gives the specific heat difference in zero applied magnetic field. dH_c/dT is a property of the material whose value does not depend on whether or not a field is actually present. Equation (5.4), on the other hand, gives the specific heat difference in the presence of an applied magnetic field in which case the temperature of the transition is reduced from T_c to T. Expressions (5.4) and (5.4a) provide a useful link between the measured critical magnetic field curve and the thermodynamic properties. We can, for example, derive the magnitude of the specific heat jump at T_c from measurements of the slope of the critical field curve. Furthermore we can provide a check on experimentally measured quantities by seeing whether they satisfy these equations.

Though there is no latent heat when, in the absence of a magnetic field, a metal undergoes the superconducting-normal transition, there is a latent heat *if a magnetic field is present*. The latent heat L for a transition between two phases a and b is given by $L = vT(s_a - s_b)$, so from (5.2) we have

$$L = -vT\mu_0 H_c \frac{dH_c}{dT}. \qquad (5.5)$$

In the absence of any magnetic field the transition occurs at the transition temperature and $H_c = 0$, but if there is a magnetic field the transition occurs at some lower temperature T where $H_c > 0$. This latent heat arises because at temperatures between T_c and 0°K the entropy of the normal state is greater than that of the superconducting state, so heat must be supplied if the transition is to take place at constant temperature. In the presence of an applied magnetic field, therefore, the superconducting-normal transition is of the *first-order*, i.e. although g is continuous, $\partial g/\partial T$ is not.

5.2.2. Adiabatic magnetization

The fact that at temperatures below T_c there is a latent heat and the entropy of the normal state is greater than that of the superconducting state has an interesting consequence. In an ordinary magnetic material application of a magnetic field decreases the entropy because the atomic dipoles become aligned in the field. This decrease in entropy with magnetic field strength is the basis of the well-known method of lowering temperature by "adiabatic demagnetization", where the temperature of a thermally isolated specimen falls as the applied magnetic field is reduced. However, the application of a sufficiently strong magnetic field to a superconducting metal will drive it into the normal state and at a given temperature this has a *greater* entropy than the superconducting state. If the specimen is thermally isolated no heat can enter and the latent heat of the transition must come from the thermal energy of the crystal lattice. Hence the temperature falls. Thus, in contrast to a paramagnetic material, a superconductor is cooled by adiabatic *magnetization*. The temperature drop to be expected can be deduced from an entropy diagram such as Fig. 5.1. If the superconductor is initially at temperature T_1, adiabatic destruction of the superconductivity by the magnetic field will take the material from point 1 to 2 and the temperature will fall to T_2.

Although Mendelssohn and his co-workers have demonstrated that a lowering of temperature can be obtained by this method, it is not used in practice to obtain very low temperatures because there are several practical disadvantages compared to other methods of cooling.

5.2.3. Lattice and electronic specific heats

There are two contributions to the specific heat of a metal. Heat raises the temperature both of the crystal lattice and of the conduction electrons. We may therefore write the specific heat of a metal as

$$C = C_{\text{latt}} + C_{\text{el}}.$$

However, as will be pointed out in Chapter 9, the properties of the lattice (crystal structure, Debye temperature, etc.) do not change when a metal becomes superconducting, and so C_{latt} must be the same in both the superconducting and normal states. Hence the difference between the specific heat values in the superconducting and normal states arises only from a change in the electronic specific heat, i.e.

$$C_s - C_n = (C_{el})_s - (C_{el})_n.$$

The fact that just below the transition temperature the specific heat is greater in the superconducting state than in the normal state implies that, when a metal in the superconducting state is cooled through this region, the entropy of its conduction electrons decreases more rapidly with temperature than if it were in the normal state [see (5.3)]. Hence on cooling a superconductor some extra form of electron order must begin to set in at the transition temperature in addition to the usual decrease in entropy of the conduction electrons which occurs when a normal metal is cooled. This additional order increases as the temperature is lowered and so gives an extra contribution to dS/dT and therefore increases the specific heat. As we have seen, the transition in zero field is second order with no latent heat and no sudden change of entropy at the transition temperature. There is at the transition temperature only a change in the *rate* at which the entropy decreases as the temperature is reduced. The two-fluid model was based on the above considerations, it being supposed that at the transition temperature some conduction electrons begin to become highly ordered superelectrons, the fraction approaching 100% as the temperature is lowered towards 0°K. The nature of this more ordered state of the electrons in a superconducting metal is discussed in Chapter 9.

Figure 5.2 shows that at temperatures well below the transition temperature the specific heat of the superconducting metal falls to a very small value, becoming even less than that of the normal metal. We have seen that the difference in the specific heats of the superconducting and normal states is the result of a change in the electronic specific heat. In order to understand the difference in specific heats we need, therefore, to be able to deduce the value of the electronic specific heat from the experimentally measured values of the total specific heat. This may be done as follows: for a *normal* metal at low temperatures the specific heat has the form

$$C_n = C_{\text{latt}} + (C_{el})_n = A\left(\frac{T}{\theta}\right)^3 + \gamma T, \qquad (5.6)$$

where A is a constant with the same value for all metals. The Debye temperature of the lattice θ, and the Sommerfeld constant γ, which is a measure of the density of the electron states at the Fermi surface, both vary from metal to metal. We can determine the lattice contribution, C_{latt}, as follows. Equation (5.6) may be re-written as

$$\frac{C_n}{T} = \left(\frac{A}{\theta^3}\right)T^2 + \gamma,$$

so a plot of the experimentally determined values of C_n/T against T^2 should give a straight line whose slope is A/θ^3 and whose intercept is γ. Hence, from measurements on the superconductor in the normal state, i.e. by applying a magnetic field greater than H_c, we can determine the lattice specific heat, $C_{\text{latt}} = A(T/\theta)^3$. The specific heat of the lattice is the same in both the superconducting and normal states, so, by subtracting the value of C_{latt} from the total specific heat C_s of the superconducting state, we can obtain the electronic contribution $(C_{\text{el}})_s$.

It is difficult to obtain accurate experimental values of the specific heats of superconductors, because at low temperatures the specific heats become very small. However, careful measurements have revealed that at temperatures well below the transition temperature the electronic specific heat of a metal in the superconducting state varies with temperature in an exponential manner,

$$(C_{\text{el}})_s = ae^{-b/kT},$$

where a and b are constants. Such an exponential variation suggests that as the temperature is raised electrons are excited across an energy gap above their ground state. The number of electrons excited across such a gap would vary exponentially as the temperature. We shall see in Chapter 9 that the BCS theory of superconductivity predicts just such a gap in the energy levels of the electrons.

The BCS theory also shows that, though the energy gap is substantially constant at very low temperatures, it decreases if the temperature is raised towards the transition temperature, falling to zero at T_c. This rapid decrease in the energy gap just below T_c accounts for the rapid increase in the specific heat of a superconducting metal as the temperature approaches T_c.

5.3. Mechanical Effects

Both the transition temperature and the critical magnetic field of a superconductor are found experimentally to be slightly altered if the material is mechanically stressed. Many of the mechanical properties of the superconducting and normal states are thermodynamically related to the free energies of these states, and we have seen that the critical magnetic field strength depends on the difference in the free energies of the two states. Hence, once it is known that the critical field changes slightly when the material is under stress, thermodynamic arguments show that the mechanical properties must be slightly different in the normal and superconducting states. For example, there is an extremely small change in volume when a normal material becomes superconducting, and the thermal expansion coefficient and bulk modulus of elasticity must also be slightly different in the superconducting and normal states. It is possible to derive expressions for these effects by straightforward thermodynamic manipulation,† but the effects are extremely small and we shall not consider them further in this book.

5.4. Thermal Conductivity

The thermal conductivity of a metal is affected if it goes into the superconducting state. Most of the heat flow along a normal metal in a temperature gradient is carried by the conduction electrons. In the superconducting state, however, the superelectrons no longer interact with the lattice in such a way that they can exchange energy, and so they cannot pick up heat from one part of a specimen and deliver it to another. Consequently, if a metal goes into the superconducting state, its thermal conductivity is reduced. This effect can be very marked at temperatures well below the critical temperature, where there are very few normal electrons left to transport the heat. For example, at 1°K the thermal conductivity of lead in the superconducting state is about 100 times less than that of the metal in the normal state.

If, however, the superconductor is driven normal by the application of a magnetic field, the thermal conductivity is restored to the higher value of the normal state. Hence the thermal conductivity of a superconductor can be controlled by means of a magnetic field, and this effect has been used in "thermal switches" at low temperatures to make and break heat contact between specimens connected by a link of superconducting metal.

---

† See, for example, E. A. Lynton, *Superconductivity*, Methuen, London, 1964.

5.5. Thermoelectric Effects

It is found, both from theory and experiment, that thermoelectric effects do not occur in a superconducting metal. For example, no current is set up around a circuit consisting of two different superconductors, if the two junctions are held at different temperatures below their transition temperatures. If a thermal e.m.f. were produced there would be a strange situation in which the current would increase to the critical value, no matter how small the temperature difference. It follows from the Thomson relations that, if there is no thermal e.m.f. in superconducting circuits, the Peltier and Thomson coefficients must be the same for all superconducting metals, and they are in fact zero.

Because all superconducting metals have identical thermoelectric constants they may, in principle, be used as a standard against which to measure other metals. The absolute values of the thermoelectric coefficients of a normal metal can be measured in a circuit comprising the metal and any superconductor.

The absence of thermoelectric effects only applies to the superconductors considered in Part I of this book. Thermoelectric effects may appear in the "Type-II" superconductors considered in Part II.

CHAPTER 6

THE INTERMEDIATE STATE

So FAR our discussion of how a magnetic field induces transitions from the superconducting to the normal state has been confined to cases in which end effects are unimportant. We ensured that this was so by considering specimens in the form of a long thin rod. We consider in this chapter what happens if we relax this restriction and consider specimens of arbitrary shape.

6.1. The Demagnetizing Factor

Consider the case of a superconducting sphere placed in a uniform magnetic field $\mathbf{H}_a$. As we have seen, the flux lines are excluded from the interior of the sphere by diamagnetic screening currents and have the form shown in Fig. 2.1 (p. 17). We shall now show that the value of the magnetic field strength inside the sphere (H_i) exceeds the value H_a which would exist if the sphere were removed.

Suppose the field $\mathbf{H}_a$ is produced by a solenoid, as shown in Fig. 6.1. One of the fundamental properties of the magnetic field vector $\mathbf{H}$ is that its line-integral around any closed curve is equal to the number of ampere-turns which link the curve (see Appendix A). Applying this result to the path $ABCDEF$ gives $\oint \mathbf{H} \cdot d\mathbf{l} = Ni$, where N is the total number of turns in the solenoid and i the current through each of them. We may write

$$\oint \mathbf{H} \cdot d\mathbf{l} = \int_{AB} \mathbf{H}_i \cdot d\mathbf{l} + \int_{BCDEFA} \mathbf{H}_e \cdot d\mathbf{l} = Ni,$$

where $\mathbf{H}_i$ is the field within the sphere and $\mathbf{H}_e$ the field at any point outside. Now if the sphere is removed the line integral is still equal to Ni, and we may write

$$\oint \mathbf{H} \cdot d\mathbf{l} = \int_{AB} \mathbf{H}_a \cdot d\mathbf{l} + \int_{BCDEFA} \mathbf{H}'_e \cdot d\mathbf{l} = Ni,$$

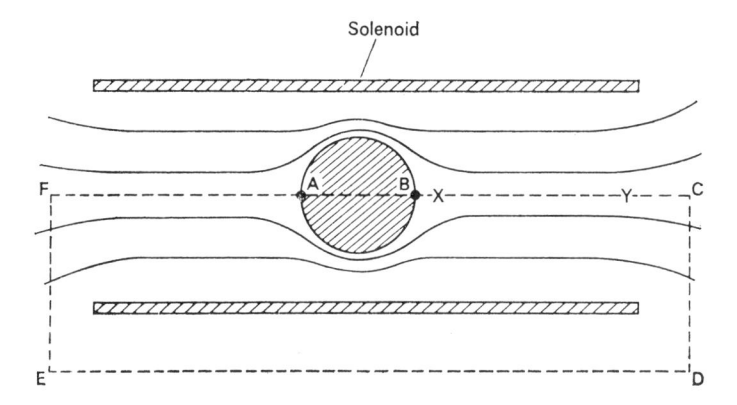

FIG. 6.1. A superconducting sphere in a solenoid. The field strength at a point close to the sphere, such as X, is less than it would be if the sphere were absent, while the field strength at a point far away, such as Y, is essentially unchanged. The line integral of H around the broken line is independent of whether the sphere is present or not, so the field strength inside the sphere must exceed the applied field H_a.

where the field between A and B in the absence of the sphere is by definition $\mathbf{H}_a$, and $\mathbf{H}'_e$ is the field at any point outside AB when the sphere is removed. Hence

$$\int_{AB} \mathbf{H}_i \cdot d\mathbf{l} + \int_{BCDEFA} H_e \cdot d\mathbf{l} = \int_{AB} \mathbf{H}_a \cdot d\mathbf{l} + \int_{BCDEFA} \mathbf{H}'_e \cdot d\mathbf{l}. \quad (6.1)$$

Now comparing H_e and H'_e at a point on the axis such as X (Fig. 6.1), H_e is clearly less than H'_e because the effect of the screening currents extends outside the sphere and distorts the flux lines (compare Fig. 2.1). But at points far from the sphere, such as Y, the presence of the sphere has a negligible effect and $H_e = H'_e$. Hence H_e is everywhere either less than or equal to H'_e, and it follows from (6.1) that H_i must be greater than H_a. In other words, although the flux density inside is zero, because of screening currents *the magnetic field strength inside the sphere exceeds the applied field strength H_a.*

This is a special case of a well-known problem in magneto-statics; namely, what is the magnetic field inside a magnetic body of arbitrary shape exposed to a uniform magnetic field? Except for the case of a long thin body or a toroid, the field inside the body differs from the applied field. In the case of a diamagnetic body such as a superconductor, the internal field exceeds the applied field, while in the case of a ferromagnetic body the internal field is less than the applied field. Because historically the study of ferromagnetism preceded that of superconductivity, the phenomenon is referred to as *demagnetization*. A magnetized body is said to produce within itself a *demagnetizing field* $\mathbf{H}_D$ which is superimposed on the applied field. So we may write the field $\mathbf{H}_i$ inside the body as $\mathbf{H}_i = \mathbf{H}_a - \mathbf{H}_D$.

For bodies of general shape the situation is complicated because the demagnetization is not uniform, the internal field varying in strength and direction throughout the body. For the special case of any ellipsoid, however, the situation is much simpler with the internal field uniform throughout the body and parallel to the applied field. The internal field is then given by

$$\mathbf{H}_i = \mathbf{H}_a - n\mathbf{I} \tag{6.2}$$

where $\mathbf{I}$ is the magnetization and n is the *demagnetizing factor* of the ellipsoid. The demagnetizing field is $n\mathbf{I}$ but, because in a superconductor $\mathbf{I}$ is negative, the internal field is increased.

Demagnetization is significant only in strongly magnetic materials because, as eqn. (6.2) shows, the difference between the strengths of the internal and applied fields is proportional to the magnetization of the body. A superconductor is, however, a strongly magnetic material; its susceptibility has a magnitude of unity, whereas the susceptibility of a typical normal metal is 10^{-4} or less. Consequently demagnetizing effects are very important in superconductors.

As we have pointed out, the demagnetizing factor n, and hence the strength of the internal field, depends on the shape of the body. The special case of a sphere is treated in most standard textbooks on electromagnetic theory.† Values of the demagnetizing factor for ellipsoids of revolution are shown in Fig. 6.2. It can be seen that for the case of a sphere $n = \frac{1}{3}$. In later chapters we shall be concerned with the effect of magnetic fields on long wires of superconductor; an infinitely long circular cylinder can be regarded as the limiting case of a long ellipsoid, so $n = \frac{1}{2}$ if H_a is perpendicular to the axis of the wire, while $n = 0$ if H_a is parallel to the axis. The case of a rod whose length is not very great compared with its diameter, in an applied field parallel to its axis, or of a flat plate perpendicular to the applied field, can be approximated quite closely by replacing the rod or plate by its inscribed ellipsoid. For the particular case of a superconductor, $I = -H_i$ and (6.2) takes the form

† For example, A. F. Kip, *Fundamentals of Electricity and Magnetism*, 2nd ed., McGraw-Hill, 1969, p. 349.

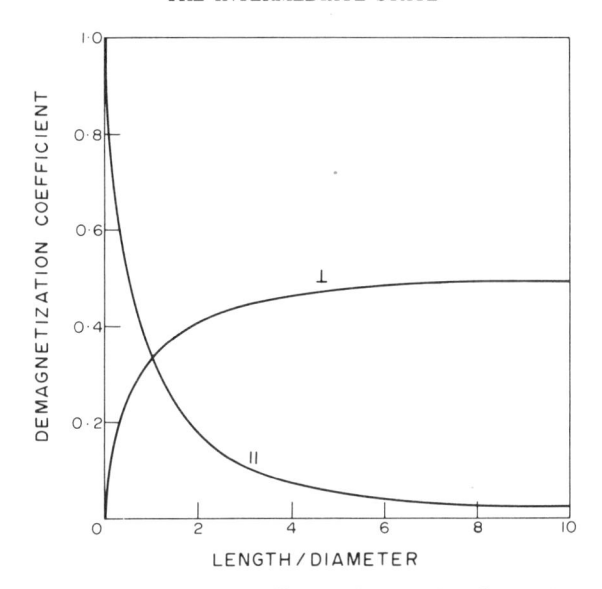

FIG. 6.2. Demagnetizing coefficient n for ellipsoids of revolution
⊥ Field perpendicular to axis of revolution
∥ Field parallel to axis of revolution

$$\mathbf{H}_i = \left(\frac{1}{1-n}\right)\mathbf{H}_a.$$

At the surface of the superconductor the tangential component of **H** is continuous, and, since the internal field is parallel to the applied field, it follows that at the equator the field strength just outside the surface is equal to the strength of the internal field H_i. Therefore, the *external* field strength at the equator exceeds the applied field strength, as can be seen from Fig. 6.1, and is given by $H_a/(1-n)$. In the case of a sphere, the external field strength at the equator is equal to $\tfrac{3}{2}H_a$, and in the case of a long cylindrical rod in a transverse field, which is a case we shall consider in Chapter 7, it has the value $2H_a$.

6.2. Magnetic Transitions for $n \neq 0$

Consider what happens to a superconducting ellipsoid when the applied field H_a is gradually raised. At first sight one might expect that when H_a reaches a value H'_c equal to $(1-n)H_c$, so that the internal field H_i becomes equal to the critical field H_c, the sphere would be driven into the normal state. But if this were to happen I would become zero, because in the normal state the susceptibility is zero, and we should have

$H_l = H_a = H_c'$, which is less than H_c. We should then have the impossible situation of a completely normal body in a field smaller than H_c. The paradox can be resolved by noting that when H_l becomes equal to H_c it is possible for the superconducting and normal phases to exist side by side in equilibrium, in the same way that a liquid can coexist with its vapour if the pressure is equal to the saturation vapour pressure. To take the simplest possible case, suppose that when H_l reaches the value H_c the ellipsoid splits up into normal and superconducting laminae parallel to the applied field as shown in Fig. 6.3. Some of the flux lines avoid the ellipsoid altogether, and others pass through the normal regions. If it

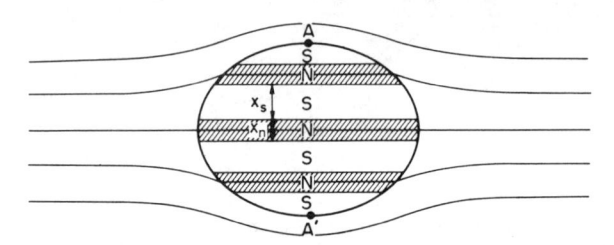

FIG. 6.3. Ellipsoid split up into normal and superconducting laminae in a magnetic field.

should turn out that the Gibbs free energy for the configuration shown in Fig. 6.3 is lower than both the free energy for the purely superconducting state and that for the purely normal state, then this arrangement of superconducting and normal domains (or something like it) will be the equilibrium state for $H_c' < H_a < H_c$. We shall see that this is indeed the case and that H_l remains equal to H_c throughout this range. Such an arrangement of normal and superconducting domains is known as the *intermediate state*, and is an essential feature of magnetic transitions for any body whose demagnetizing factor is not zero. The model we have adopted here, in which the normal and superconducting regions are plane parallel laminae, is an oversimplified model, and in general the way in which the body splits up is very complicated. Nevertheless, this simple model brings out surprisingly well most of the important features of the intermediate state.

6.3. The Boundary Between a Superconducting and a Normal Region

Before proceeding to discuss the intermediate state in some detail, we first consider the conditions which must exist if there is to be a stationary boundary between a superconducting and a normal region. Suppose the field strength in the normal region is H, so that the flux density B in the normal region is equal to $\mu_0 H$. A fundamental property of the magnetic flux density vector **B** is that its component perpendicular to any boundary between two media is continuous, and the two phases are here playing the role of two different media as far as their magnetic properties are concerned. In the superconducting region **B** is everywhere zero, so it follows that there can be no component of **B** perpendicular to the boundary in the normal region and that on the normal side **B** and **H** must both lie parallel to the boundary. In other words, boundaries between normal and superconducting regions must lie parallel to the local direction of the magnetic field.

There is also an important restriction on the value of the magnetic field strength at the boundary. The component of **H** parallel to any boundary between two different magnetic media must be continuous, and we have shown that **H** is in fact parallel to the boundary on the normal side. It follows that the magnitude of **H** must be the same on both sides. At the boundary the magnitude of **H** must equal H_c. If it were less, the material on the normal side could lower its free energy by becoming superconducting. On the other hand, the field strength cannot exceed H_c or the material on the superconducting side of the boundary would be driven normal. We see, therefore, that a stationary boundary will only exist where the field strength is exactly H_c, and where there is such a stationary boundary the flux density on the normal side will be $\mu_0 H_c$. (It will appear later on that this restriction on the value of H at the boundary has to be modified slightly if there is a "surface energy" associated with the interface between the normal and superconducting regions.)

6.4. Magnetic Properties of the Intermediate State

We shall now discuss the magnetic properties of a body split up into normal and superconducting regions as in Fig. 6.3. First of all we need to know the effective value of B inside such a body, which we will assume to be an ellipsoid so that we can assign to it a demagnetizing factor n. The effective value of B is the flux density averaged over a region whose dimensions are large compared with the cross-sections of the normal and superconducting laminae. In other words, the effective value of B, which we write as $\bar{B}$, is simply the total flux passing through the ellipsoid divided by its maximum cross-sectional area. Since the flux density in the superconducting regions is zero, $\bar{B}$ is simply ηB_n, where B_n is the local flux density in the normal regions and η is the fraction of the cross-section which is in the normal state. From Fig. 6.3, $\eta = x_n/(x_n + x_s)$. Also, in the normal regions $B_n = \mu_0 H_i$, where H_i is the internal field, so that $\overline{B} = \eta \mu_0 H_i$. If we regard the ellipsoid as having an effective relative permeability $\bar{\mu}_r$ such that $\overline{B} = \bar{\mu}_r \mu_0 H_i$, then $\bar{\mu}_r = \eta$. Also, we can assign to it an average magnetization $\overline{I} = \left(\dfrac{\overline{B}}{\mu_0} - H_i\right)$.

Now
$$H_i = H_a - n\overline{I}$$
$$= H_a - n\left(\frac{\overline{B}}{\mu_0} - H_i\right)$$
$$= H_a - n(\eta - 1)H_i$$

or
$$H_i = H_a/[1 + n(\eta - 1)]. \tag{6.2}$$

We have already pointed out that the normal and superconducting phases can coexist only if the magnetic field is equal to H_c, so we shall assume that the body splits up in such a way that $H_i = H_c$. If $H_i = H_c$, then (6.2) gives η for any particular value of H_a between H_c' and H_c. For all values of H_a in this range, the value of η so obtained gives a value of $\overline{B}$, and hence of $\overline{I}$, which is just right to make H_i equal H_c, so the assumption that $H_i = H_c$ is self-consistent.

It is instructive to plot the variables H_i, η, $\overline{B}$ and $\overline{I}$ as functions of the applied field strength H_a. These are shown in Fig. 6.4a–d. The graph (d), showing $\overline{I}$, is particularly important in practice because the total magnetic moment of the body, which is often measured experimentally, is given by $V\overline{I}$.

6.5. The Gibbs Free Energy in the Intermediate State

To find the Gibbs function G in the intermediate state we start with $H_a = 0$, so that $G(0) = Vg_s(0)$, and make use of the result $dG = -\mu_0 M dH_a$, where $M = V\overline{I}$ is the total magnetization of the body.[†] Hence,
$$G(H_a) = G(0) - \int_0^{H_a} \mu_0 M\, dH_a.$$

There are three distinct regions of integration to be considered:

(i) $0 < H_a < (1 - n)H_c$

[†] Note that the expression for dG involves dH_a, not dH_i. See A. B. Pippard, *Elements of Classical Thermodynamics*, C.U.P., 1961, p. 26.

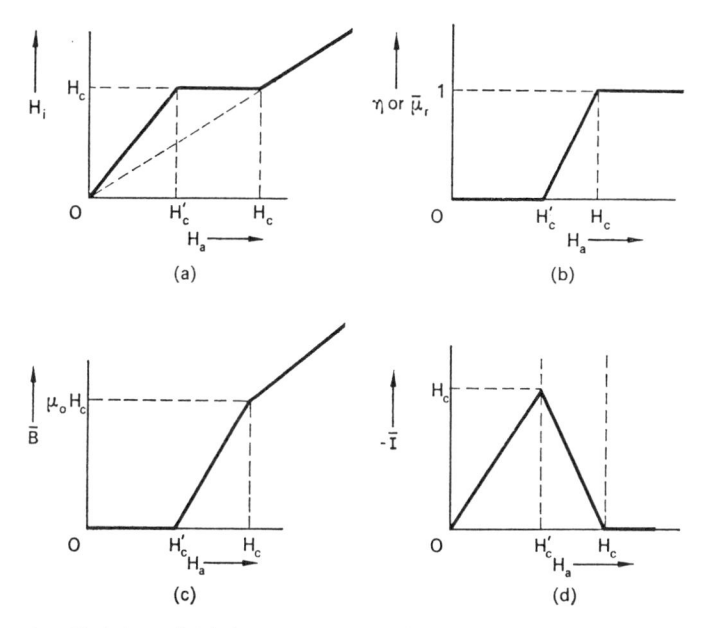

FIG. 6.4. Variation of (a) internal magnetic field strength, (b) effective relative permeability, (c) effective flux density, and (d) intensity of magnetization with applied magnetic field.

In this range of field strengths the sphere is totally superconducting and $I = -H_i$.

Hence
$$G(H_a) = Vg_s(0) + \frac{V\mu_0 H_a^2}{2(1-n)}.$$

(ii) $(1-n)H_c < H_a < H_c$

The sphere is in the intermediate state and $\bar{I} = \dfrac{\bar{B}}{\mu_0} - H_i$.

Therefore $\bar{I} = (\eta - 1)H_i = (\eta - 1)H_c$.
In this range
$$G(H_a) = Vg_s(0) + \frac{V\mu_0 H_c}{2n}\left[H_a\left(2 - \frac{H_a}{H_c}\right) - H_c(1-n)\right]$$

(iii) $H_a > H_c$
The sphere is completely normal and $\bar{I} = 0$.
Here $G(H_a) = Vg_s(0) + \tfrac{1}{2}V\mu_0 H_c^2 = Vg_n(0)$.

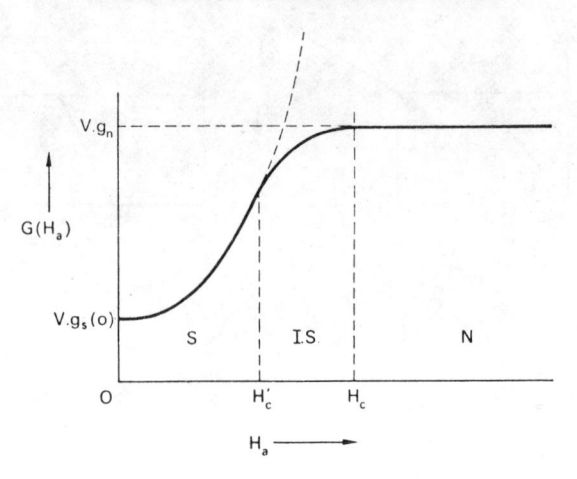

FIG. 6.5. Variation of Gibbs free energy with H_a for a body with non-zero demagnetizing factor.

The variation of G with H_a is shown in Fig. 6.5, which demonstrates clearly that the intermediate state (as represented by the simple model of plane laminae) has a lower free energy than either the purely superconducting or the purely normal state for $H'_c < H_a < H_c$.

It should be noted that since the component of the magnetic field parallel to the boundary between two media is the same on each side of the boundary, and since in addition H_l is uniform and equal in magnitude to H_c for $(1 - n) H_c < H_a < H_c$, the *external* field at the equator (AA' in Fig. 6.3) is equal to H_c whenever the sphere is in the intermediate state. Some authors introduce the intermediate state by saying that the specimen must leave the purely superconducting state when the *external* field strength at the equator is equal to H_c. This involves a different approach from the one we have used, and we prefer to adopt the standpoint of finding the state with the lowest free energy.

6.6. The Experimental Observation of the Intermediate State

The domain structure in the intermediate state has been revealed experimentally by many workers. The first of these were Meshkovsky and Shalnikov, who explored the magnetic field in the narrow gap between two hemispheres of tin of diameter 4 cm with a very small bismuth probe. (This relies on the fact that the resistivity of bismuth is very sensitive to the presence of a magnetic field.) The bismuth probe measures the local value of B in the region of the hemisphere immediate-

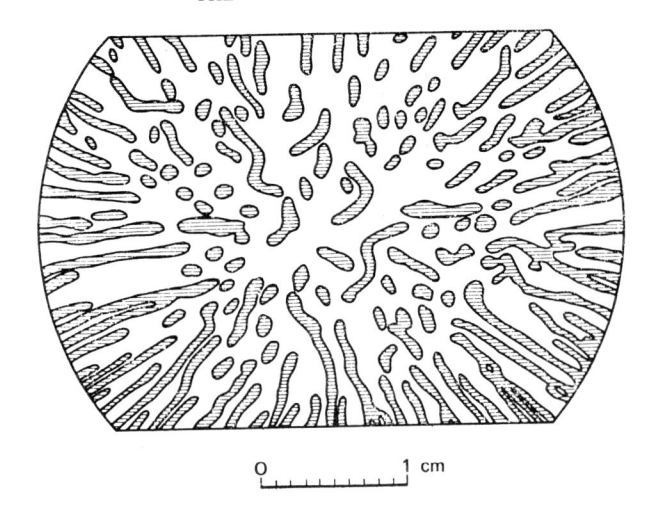

FIG. 6.6. Structure of intermediate state in tin sphere, after Meshkovsky ($T = 2\cdot 85°$K, $H_a = 0\cdot 7\, H_c$). The shaded areas represent normal regions.

ly adjacent to it, which is zero in the superconducting domains and equal to $\mu_0 H_c$ in the normal regions. A map showing their results is shown in Fig. 6.6, where the normal regions are shown shaded. The normal regions appear to consist partly of radial laminae and partly of roughly cylindrical filaments, showing, as might be expected, that the laminar model adopted in § 6.3 is too simple.

Other methods of observing the intermediate state rely on the tendency of ferromagnetic powders to accumulate in regions of high flux density, of superconducting (diamagnetic) powders to accumulate in regions of low flux density, or on the ability of a paramagnetic glass in a magnetic field to rotate the plane of polarization of polarized light (Faraday effect). Most observations of the intermediate state have been carried out with flat plates which have a value of n very close to unity for a perpendicular field. By approximating a circular plate of radius a and thickness t by its inscribed ellipsoid, it can be shown that $n \cong 1 - (t/2a)$ so that $H_c(1 - n)$ is a very low field, and almost any value of H_a is sufficient to drive the plate into the intermediate state. A beautiful photograph of the intermediate state in aluminium plates, obtained by Faber, is shown in Fig. 6.7. This was taken by dusting the plates with tin powder so that the superconducting tin tended to accumulate in regions of low magnetic field strength, i.e. adjacent to the superconducting regions of the aluminium. The normal regions are seen to be laminar, but the laminae are not flat and have a characteristic structure.

74 INTRODUCTION TO SUPERCONDUCTIVITY

Fig. 6.7. Intermediate state in aluminium plate 0·47 cm thick with magnetic field perpendicular to surface ($H = 0·65\,H_c$, $T = 0·92\,T_c$). The dark lines are covered with tin powder and correspond to superconducting regions (after Faber).

6.7. The Absolute Size of the Domains: the Role of Surface Energy

The simple analysis given in § 6.4 shows how the ratio of the widths of the superconducting and normal regions is related to the applied magnetic field, but it tells us nothing about their absolute values. On this simple model the absolute values are indeterminate, since the Gibbs function depends only on the total amount of normal or superconducting material, i.e. on the ratio x_s/x_n. To discuss the factors that determine the absolute values of x_s and x_n we need to introduce a new concept, that of a possible *surface energy* between the normal and superconducting phases; in other words, we consider the possibility that extra energy may have to be supplied to form a boundary between the two phases. Such a concept is quite common in physics; the surface tension which exists at the boundary between a liquid and its vapour is an obvious example, and the importance of surface tension in controlling the equilibrium size of small droplets is well known. We shall for the moment assume that such a surface energy exists, and defer until § 6.9 the question of how it arises. It is easy to see qualitatively what the role of such a surface energy may be in controlling the structure of the intermediate state. The surface energy will give to the Gibbs free energy an additional contribution which is proportional to the total area of the boundary between the normal and superconducting phases. If it is positive, then the free energy is minimized by having as small an area of interphase boundary as possible, i.e. by having a few thick domains; on the other hand, if the surface energy is negative, i.e. if energy is released on forming the interphase boundary, it is energetically favourable for the body to split up into a large number of thin domains so as to make the area of the interphase boundary as large as possible. In fact, if the surface energy is negative, a superconductor in a magnetic field would tend to split up into normal and superconducting regions even in the absence of demagnetizing effects, and not show a true Meissner effect. It was the occurrence of the Meissner effect in pure superconductors such as lead or tin that led London to infer that in these superconductors there must be a positive surface energy.

The energy per unit area of boundary between the superconducting and normal regions is usually denoted by α, and it often simplifies the mathematical analysis if the surface energy is expressed in terms of a characteristic length Δ such that $\alpha = \tfrac{1}{2}\mu_0 H_c^2 \Delta$. An exact analysis of the domain structure in the intermediate state is a formidable theoretical problem, and can only be attempted if some assumption about the shape of the domains is made. For the case of a flat plate in a perpendicular field, the problem has been analysed by Landau and by Kuper on the assumption that the domains are laminar; both find that the ratio of the thickness of the superconducting regions x_s to the thickness of the plate d is of the order of $(\Delta/d)^{\frac{1}{2}}/h$, where $h = H_a/H_c$. In practice, for superconductors such as lead or tin, Δ is found to be about 5×10^{-5} cm, so that for a plate of 1 cm thickness in a perpendicular magnetic field whose value is near to the critical field, $(x_s/d) \sim 10^{-2}$, or $x_s \sim 10^{-2}$ cm. If the field strength is only about one-tenth of the critical field, then $x_s \sim 10^{-1}$ cm. Experimental observation of the domain size in the intermediate state provided one of the earliest ways of measuring the surface energy, but owing to the uncertainty in the theory the accuracy is not high.

6.8 Restoration of Resistance to a Wire in a Transverse Magnetic Field

We have seen in Chapter 4 that the restoration of resistance to a long superconducting cylinder in an axial magnetic field occurs abruptly when the applied field reaches the value H_c, and that in favourable circumstances the transition can be very sharp. If the field is applied perpendicular to the axis, the variation of resistance with field is quite different and is typically as shown in Fig. 6.8.

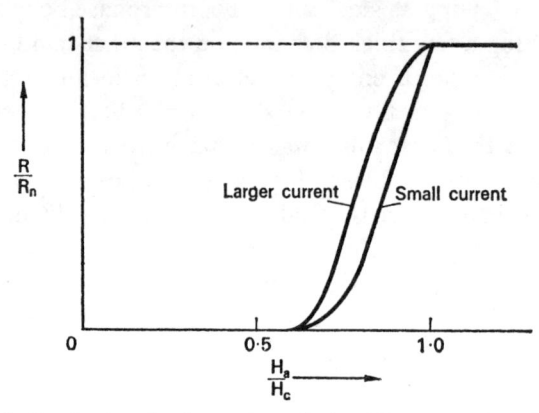

FIG. 6.8. Restoration of resistance to a wire by a transverse magnetic field.

There are three essential features:

(i) The resistance begins to return when H_a is little more than $0.5\ H_c$. This is understandable because the wire has been driven into the intermediate state, which for a cylinder should happen for fields greater than $0.5\ H_c$. We have to assume that the domains have the approximate form of laminae perpendicular to the axis, so that there is no continuous superconducting path from one end of the wire to the other. The fact that the onset of resistance invariably occurs slightly above the value $0.5\ H_c$ which we should expect for a circular cylinder can be understood in terms of an interphase surface energy; the analysis of § 6.5, which neglects surface energy, predicts that the intermediate state should set in at $H_a = (1 - n)\ H_c$. If there is a positive surface energy, there will be an additional contribution to the Gibbs free energy in the intermediate state, and a value of H_a greater than $(1 - n)\ H_c$ must be reached before the free energy of the intermediate state is lower than that of the pure superconducting state.†

† Since, for applied field strengths greater than $(1 - n)H_c$, the internal field strength exceeds H_c, there appears to be a contradiction here with the statement on p. 69 that at a normal-superconducting boundary the local value of the magnetic field strength is always equal to H_c. This contradiction can be resolved by noting that the presence of surface energy modifies the value which the local magnetic field strength must have at a stationary boundary. The effect is analogous to the increase in the pressure of a vapour in equilibrium with a small drop of liquid

(ii) The resistance increases smoothly with H_a, reaching the full normal resistance when H_a attains the value H_c. This again can be understood if the laminae are perpendicular to the axis, since in the intermediate state η is a linear function of H_a.

(iii) The exact shape of the R versus H curve depends on the magnitude of the measuring current. The implication of this is that the simple laminar model is too crude; the superconducting regions tend to be linked by superconducting filaments which can only carry a small current. As the current increases, these filaments are driven normal and the resistance increases.

6.9. The Concept of Coherence and the Origin of the Surface Energy

To explain the origin of surface energy at a boundary between normal and superconducting regions, it is necessary to introduce a very important concept formulated by Pippard[†] in 1953—that of the *coherence length*. To explain coherence we must first return to the idea of the superconducting state as a highly ordered state. We saw in § 5.2.3 that when a superconductor is cooled below the transition temperature some extra form of order sets in amongst the conduction electrons, and in § 1.4 we introduced the idea that a superconductor can be regarded as consisting of two interpenetrating electronic fluids, the normal electrons and the superelectrons. The superelectrons in some way possess greater order than the normal electrons, and we can think of the degree of order of the superconducting phase as being identified with the density of superconducting electrons n_s. By considering several aspects of the behaviour of superconductors Pippard was led to the idea that n_s cannot change rapidly with position, but can only change appreciably within a

---

when there is surface tension at the liquid–vapour boundary. If the surface energy is positive, the equilibrium value of the magnetic field strength at a superconducting-normal boundary is given by

$$H = H_c \left(1 + \frac{\Delta}{2r}\right),$$

where Δ is defined on p. 75 by $\alpha = \frac{1}{2}\mu_0 H_c^2 \Delta$, and r is the radius of curvature of the boundary. If the normal regions were plane laminae, r would be infinite and we should have $H = H_c$. In practice the domains are never plane laminae but always possess some curvature, so that $H > H_c$ if Δ is positive.

[†] A. B. Pippard, *Physica* 19, 765 (1953).

distance which for pure superconductors is of the order of 10^{-4} cm. This distance he called the coherence length ξ. One consequence of the existence of the coherence length is that the boundary between a normal and superconducting region cannot be sharp because the density of superelectrons can rise from zero in the normal region to its density n_s in the superconducting region only gradually over a distance equal to about the coherence length (see dotted line in Fig. 6.9a).

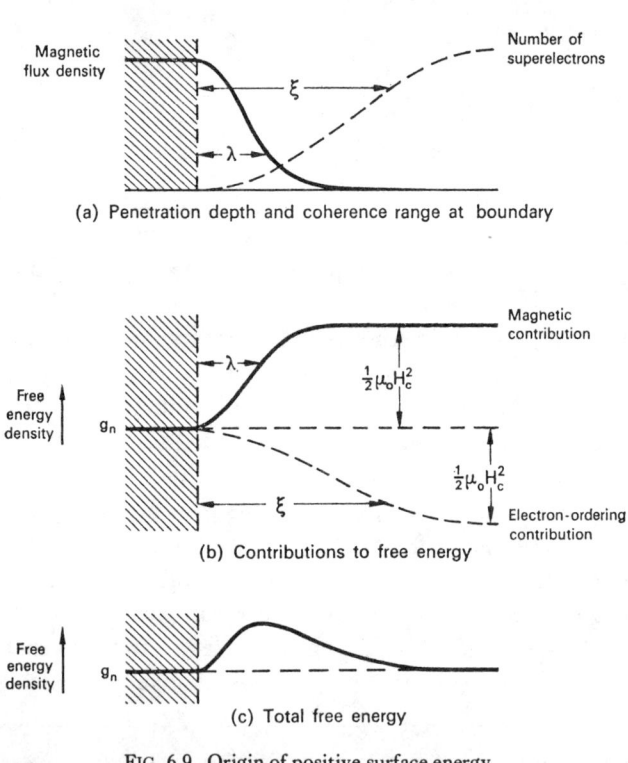

FIG. 6.9. Origin of positive surface energy.

An important property of the coherence length is that it depends on the purity of the metal, the figure of 10^{-3} mm which we have quoted being representative of a pure superconductor. If impurities are present the coherence length is reduced. The coherence length in a perfectly pure superconductor, which is an intrinsic property of the metal, is usually denoted by ξ_0, while the actual coherence length in an impure metal or alloy is written as ξ. In very impure specimens, which are characterized by a very short electron mean free path l_e, the coherence length is reduced to approximately $(\xi_o l_e)^{\frac{1}{2}}$.

There are several arguments, mostly of a circumstantial nature, which lead to the concept of coherence. Probably the simplest and most direct arises from the extreme sharpness of the transition in zero field. In a pure well-annealed specimen the transition may be less than 10^{-5} degrees wide, and this suggests that the co-operation of a very large number of electrons is involved, as otherwise statistical fluctuations from point to point would broaden the transition. The idea of coherence as we have introduced it here may seem rather vague and ill defined, as indeed it was at its inception, but we shall see when we discuss the microscopic theory of superconductivity in Chapter 9 that the concept can be put on a more precise and quantitative basis. It is also in line with the predictions of the Ginzburg–Landau theory discussed in § 8.5.

Another argument in favour of the idea of coherence is that it allows a simple explanation of the origin of surface energy, as we shall now show. Consider a superconducting region adjoining a normal region. As we saw in § 6.3, this situation can only arise in the presence of a magnetic field of strength H_c. At the boundary, there is not a sudden change from fully normal behaviour to fully superconducting behaviour; the flux density penetrates a distance λ into the superconducting region, and in accordance with the coherence concept, in the superconducting region the number of superelectrons per unit volume n_s increases slowly over a distance about equal to the coherence length ξ (Fig. 6.9a).

Now consider the free energy at the boundary. If the boundary is to be stable the superconducting and normal regions must be in equilibrium, that is to say their free energy per unit volume must be the same. There are two contributions which change the free energy of the superconducting region relative to that of the normal region. Due to the presence of the ordered superelectrons the free energy density of the superconducting state is lowered by an amount $g_n - g_s$, and, in addition, because the superconducting region has acquired a magnetization which cancels the flux density inside, there is a positive "magnetic" contribution $\frac{1}{2}\mu_0 H_c^2$ to its free energy density. For equilibrium $\frac{1}{2}\mu_0 H_c^2 = g_n - g_s$, so that well inside the superconducting region the two contributions cancel and the free energy density is the same as in the neighbouring normal region. At the boundary, however, the degree of order (i.e. the number of superelectrons n_s) rises only gradually over a distance determined by the coherence length ξ, so the decrease in free energy due to the increasing order of the electrons takes place over the same distance (Fig. 6.9b). On the other hand the "magnetic" contribution to the free energy rises over a distance of about the penetration depth λ. In general ξ and λ are not the same, so the two contributions do not cancel near the boundary. If, as in Fig. 6.9, the coherence length is *longer* than the penetration depth, the total free energy density is increased close to the boundary; that is to say, there is a positive surface energy. It can be seen from Fig. 6.9b that, roughly speaking, the value of this surface energy is approximately $\frac{1}{2}\mu_0 H_c^2(\xi - \lambda)$ per unit area of the boundary. (This can be seen by replacing the two curves in Fig. 6.9a by rectangular steps in which the changes in flux density and n_s occur abruptly at distances λ and ξ respectively from the edge of the normal region.) The length Δ introduced in § 6.7 is therefore to be identified with $\xi - \lambda$, and since the domain size in the intermediate state enables Δ to be measured, this gives the value of $\xi - \lambda$. As we saw in § 6.7, the value of Δ in a typical pure superconductor such as lead or tin is about 5×10^{-5} cm, which is about 10 times larger than λ, so it is clear that in this case $\xi \simeq \Delta \simeq 5 \times 10^{-5}$ cm. This is, in fact, one of the arguments which enabled Pippard to assign an order of magnitude of ξ_0.

Coherence is a very fundamental property of superconductors; for example, we shall see that the coherence length plays an important role in determining the properties of the type-II superconductors which we consider in the second part of this book.

The reader should be warned that in the literature on superconductivity the term "coherence length" is used in two rather different senses. The coherence length which we have been discussing, and which is of concern to us in this book, should strictly be called the "temperature-dependent coherence length", $\xi(T)$. As we have seen, it is related to the fact that the superconducting order parameter cannot vary rapidly with position; i.e. the density of superelectrons, or the energy gap or the amplitude of the electron-pair wave can only change by a significant amount over a distance greater than $\xi(T)$. This accounts for the lack of sharpness of the boundary between superconducting and normal regions. This coherence range, which is a special feature of a superconductor, varies with temperature, increasing at higher temperature in much the same way as the penetration depth (see § 2.4.1).

The other coherence length, the "temperature-independent coherence length", is related to the correlation between the motion of electrons. Because the motion of one electron is correlated with the motion of other electrons which may be some distance away, the current density at any point is not determined just by the fields at that point but by the fields averaged over a volume surrounding that point. The radius of this volume over which the electrons experience the average field is the temperature-independent coherence range. Strictly speaking, this coherence range is not peculiar to superconductors; there is some correlation of the electron motion in normal conductors, where the coherence length is the electron mean free path. In this book, however, we are only concerned with the temperature-dependent coherence length associated with the slowness with which superconducting properties can change with position.

CHAPTER 7

TRANSPORT CURRENTS IN SUPERCONDUCTORS

7.1. Critical Currents

THE EARLY workers in superconductivity soon discovered that there is an upper limit to the amount of current that can be passed along a piece of superconductor if it is to remain resistanceless. We call this the *critical current* of that particular piece. If the current exceeds this critical value, some resistance appears.

We now show that the critical current is related to the critical magnetic field strength H_c. We saw in Chapter 3 that all currents in a superconductor flow at the surface within the penetration depth, the current density decreasing rapidly from some value $\mathcal{J}_a$ at the surface. It was pointed out in Chapter 4 that superconductivity breaks down if the supercurrent density exceeds a certain value which we call the critical current density $\mathcal{J}_c$.

In general there can be two contributions to the current flowing on the surface of a superconductor. Consider, for example, a superconducting wire along which we are passing a current from some external source such as a battery. We call this current the "transport current" because it transports charge into and out of the wire. If the wire is in an applied magnetic field, screening currents circulate so as to cancel the flux inside the metal. These screening currents are superimposed on the transport current, and at any point the current density $\mathbf{J}$ can be considered to be the sum of a component $\mathbf{J}_i$ due to the transport current and a component $\mathbf{J}_H$ which arises from the screening currents

$$\mathbf{J} = \mathbf{J}_i + \mathbf{J}_H.$$

We may expect that superconductivity will break down if the magnitude of the *total* current density $\mathbf{J}$ at any point exceeds the critical current density $\mathcal{J}_c$.

According to the London equation (3.17) there is a relation between the supercurrent density at any point and the magnetic flux density at that point, and this same relation holds whether the supercurrent is a screening current, a transport current or a combination of both. Hence, when a current flows on a superconductor, there will at the surface be a flux density B and a corresponding field strength $H(=B/\mu_0)$ which is related to the surface current density $\mathcal{J}_a$.

If the total current flowing on a superconductor is sufficiently large, the current density at the surface will reach the critical value $\mathcal{J}_c$ and the associated magnetic field strength at the surface will have a value H_c. Conversely, a magnetic field of strength H_c at the surface is always associated with a surface supercurrent density $\mathcal{J}_c$. This leads to the following general hypothesis: *a superconductor loses its zero resistance when, at any point on the surface, the total magnetic field strength, due to transport current and applied magnetic field, exceeds the critical field strength H_c.* The maximum amount of *transport* current which can be passed along a piece of superconductor without resistance appearing is what we call the critical current of that piece. Clearly the stronger the applied magnetic field the smaller is this critical current.

If there is no applied magnetic field the only magnetic field will be that generated by any transport current, so in this case, the critical current will be that current which generates the critical magnetic field strength H_c at the surface of the conductor. This special case of the general rule stated above is known as Silsbee's hypothesis[†] and was formulated before the concept of critical current density was appreciated. We shall call the more general rule for the critical current given in the previous paragraph the "generalized form" of Silsbee's hypothesis.

We saw in Chapter 4 that the critical magnetic field strength H_c depends on the temperature, decreasing as the temperature is raised and falling to zero at the transition temperature T_c. This implies that the critical current density depends on temperature in a similar manner, the critical current density decreasing at higher temperatures. Conversely, if a superconductor is carrying a current, its transition temperature is lowered.

7.1.1. Critical currents of wires

Let us consider a cylindrical wire of radius a. If, in the absence of any externally applied magnetic field, a current i is passed along the wire, a magnetic field will be generated at the surface whose strength H_i is given by

[†] F. B. Silsbee, *J. Wash. Acad. Sci.*, **6**, 597 (1916).

$$2\pi a H_i = i.$$

The critical current will therefore be

$$i_c = 2\pi a H_c. \tag{7.1}$$

This relation for the critical current can be tested by measuring the maximum current a superconducting wire can carry without resistance appearing, and it is found that, in the absence of any externally applied magnetic field, eqn. (7.1) predicts the correct value.

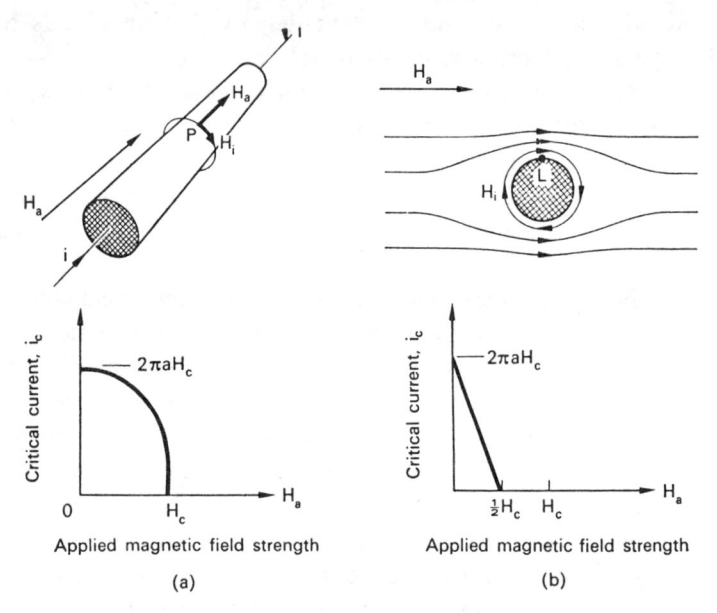

FIG. 7.1. Variation of critical current with applied magnetic field strength. (a) Longitudinal applied field. (b) Transverse applied field (transport current flowing into page).

In zero or weak applied magnetic field strengths the critical currents of superconductors can be quite high. As an example, consider a 1 mm diameter wire of lead cooled to 4·2°K by immersion in liquid helium. At this temperature the critical field of lead is about $4 \cdot 4 \times 10^4$ A m^{-1} (550 gauss) so, in the absence of any applied magnetic field, the wire can carry up to 140 A of resistanceless current.

Let us now consider to what extent the critical current is reduced by the presence of an externally applied magnetic field. First suppose that an applied magnetic field of flux density B_a and strength H_a ($= B_a/\mu_0$) is in a direction parallel to the axis of the wire (Fig. 7.1a). If a current i is passed along the wire it generates a field encircling the wire, and at the surface of the wire the strength of this field is $H_i = i/2\pi a$. This field and the applied field add vectorially and, because in this case they are at right angles, the strength H of the resultant field at the surface is given by $(H_a^2 + H_i^2)^{\frac{1}{2}}$ or

$$H^2 = H_a^2 + (i/2\pi a)^2.$$

The critical value i_c of the current occurs when H equals H_c:

$$H_c^2 = H_a^2 + \frac{i_c^2}{4\pi^2 a^2}. \tag{7.2}$$

H_c is a constant, and so this equation, which expresses the variation of i_c with H_a, is the equation of an ellipse. Consequently, the graph representing the decrease in critical current as the strength of a longitudinal applied magnetic field is increased has the form of a quadrant of an ellipse (Fig. 7.1a). In this configuration the magnetic flux density is uniform over the surface of the wire and the flux lines follow helical paths.

Another case of importance occurs when an applied magnetic field is normal to the axis of the wire (Fig. 7.1b). (We assume here that the applied field is not strong enough to drive the superconductor into the intermediate state.) In this case the total flux density is not uniform over the surface of the wire; the flux densities add on one side of the wire and substract on the other. The maximum field strength occurs along the line L. Here, because of demagnetization, as pointed out at the end of § 6.1, a field $2H_a$ is superimposed on the field H_i to give a total field

$$H = 2H_a + H_i = 2H_a + \frac{i}{2\pi a}.$$

The general form of Silsbee's rule states that resistance first appears when the total magnetic field strength at *any* part of the surface equals H_c, and so the critical current in this case is given by

$$i_c = 2\pi a(H_c - 2H_a). \tag{7.3}$$

In this case, therefore, the critical current decreases linearly with increase in applied field strength, falling to zero at $\frac{1}{2}H_c$.

It should be emphasized that the critical current of a specimen is defined as the current at which it ceases to have zero resistance, *not* as the current at which the full normal resistance is restored. The amount of resistance which appears when the critical current is exceeded depends on a number of circumstances, which we examine in the next section.

7.2. Thermal Propagation

The variation of critical current with applied magnetic field predicted by (7.2) and (7.3) has been confirmed by experiment, though measurement of critical currents, especially in low magnetic fields where the current values can be high, is not always easy. To see why there may be a difficulty, we now examine the processes by which resistance returns to a wire when the critical current is exceeded. Consider, for example, a cylindrical wire of superconductor. In practice no piece of wire can have perfectly uniform properties along its length; there may be accidental variations in composition, thickness, etc., or the temperature may be slightly higher at some points than others. As a result the value of critical

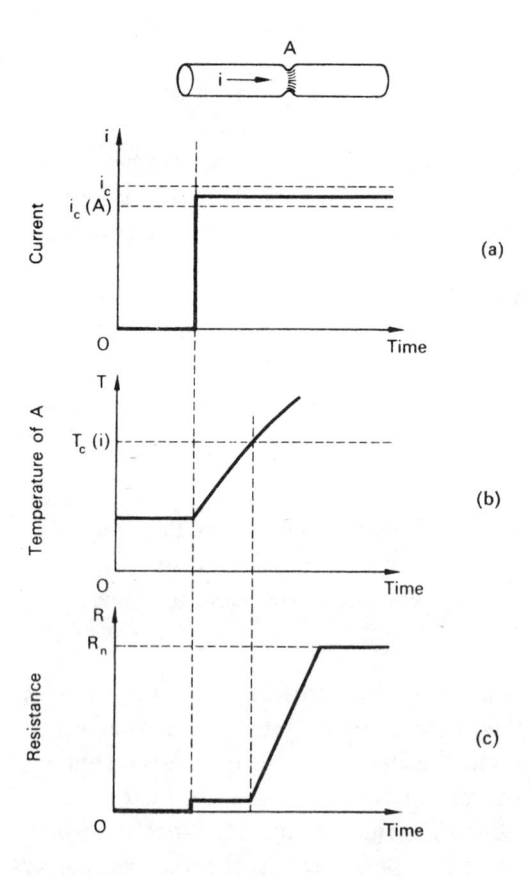

FIG. 7.2. Thermal propagation. (b) shows the temperature variation of the region A resulting from the current increase shown at (a). (c) shows the return of the wire's resistance when the normal region spreads from A.

current will vary slightly from point to point, and there will be some point on the wire which has a lower critical current than the rest of the wire. In Fig. 7.2 such a region is represented by the section A where the wire is slightly narrower. Suppose we now pass a current along the wire and increase its magnitude, until the current just exceeds the critical current $i_c(A)$ of the section A (Fig. 7.2a). This small section will become resistive while the rest of the wire remains superconducting, so a very small resistance r appears in the wire. At A the current i is flowing through resistive material and at this point heat is generated at a rate proportional to $i^2 r$. Consequently the temperature at A rises, and heat flows away from A along the metal and into the surrounding medium at a rate which depends on the temperature increase of A, the thermal conductivity of the metal, the rate of heat loss across the surface, etc. The temperature of A will rise until the rate at which heat flows away from it equals the rate $i^2 r$ at which the heat is generated. If the rate of heat generation is low, the temperature of A rises only a small amount and the wire remains indefinitely in this condition. If, however, heat is generated at a high rate, either because the resistance of A is high or because the current i is large, the temperature of A may rise above the critical temperature of the wire (Fig. 7.2b). The presence of the current has in fact reduced the transition temperature of the superconducting wire from T_c to a lower value $T_c(i)$, and if, as a result of the heating of A, the regions adjacent to A are heated above $T_c(i)$ they will become normal. The current i is now flowing through these new normal regions and generates heat which drives the regions adjacent to them normal. Consequently, even though the current i is held constant, a normal region spreads out from A until the whole wire is normal and the full normal resistance R_n is restored (Fig. 7.2c). This process whereby a normal region may spread out from a resistive nucleus is called *thermal propagation*, and we see that it is more likely to occur if the critical current is large and if the resistivity of the normal state is high (e.g. in alloys).

On account of thermal propagation there can be difficulty in measuring the critical current of a specimen, especially in low or zero magnetic fields where the current value may be high. Consider a superconducting specimen of uniform thickness, as shown in Fig. 7.3a, whose critical current we are attempting to measure by increasing the current until a voltage is observed. If the current is less than the critical current there will be no voltage drop along the specimen and no heat will be generated in it. However, the leads carrying the current to the specimen are usually of ordinary non-superconducting metal and so heat is generated in these by the passage of the current. Consequently the ends of the specimen, where it makes contact with the leads, will be slightly heated and will have a lower critical current than the body of the specimen. As the current is increased, therefore, the ends go normal at a current less than the true critical current of the specimen, and normal regions may spread through the wire by thermal propagation. Consequently a voltage is observed at a current less than the true critical value. To lessen the risk of thermal propagation from the contacts one

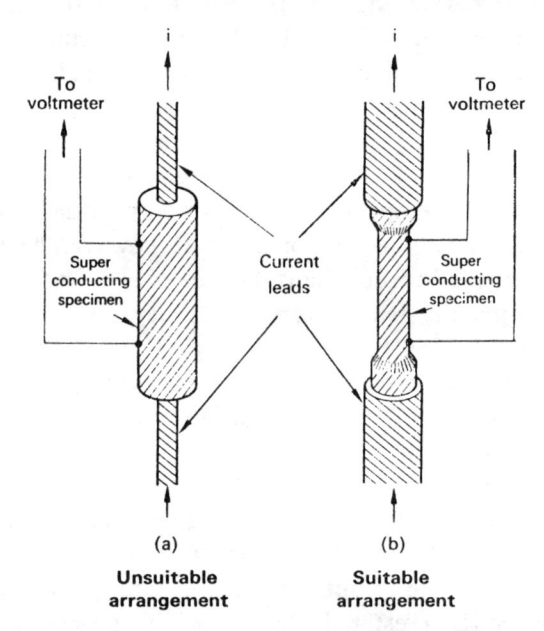

FIG. 7.3. Measurement of critical current.

should use as thick current leads as possible so that little heat is produced in them. It is also desirable to make the ends of the superconducting specimen thicker than the section whose critical current we are measuring, so that the critical current of the thinner section will be reached before thermal propagation starts from the ends (Fig. 7.3b).

A characteristic of the return of resistance by thermal propagation is the complete appearance of the full normal resistance once a certain current has been exceeded, as a result of the normal region spreading right through the specimen.

7.3. Intermediate State Induced by a Current

If thermal propagation does not occur, the full normal resistance does not appear at a sharply defined value of current but over a considerable current range. Consider a cylindrical wire of superconductor with a critical field strength H_c. If the radius of the wire is a, a current i produces a magnetic field strength $i/2\pi a$ at the surface. As we have seen, the greatest current the wire can carry while remaining wholly superconducting must be $i_c = 2\pi a H_c$, because, if the current were to exceed this, the magnetic field strength at the surface would be greater than H_c.

We might at first suppose that at i_c an outer cylindrical sheath is driven normal while the centre remains superconducting. However, this is not possible, as we shall now show. Suppose that an outer sheath becomes normal, leaving a cylindrical core of radius r in the superconducting state. The current will now flow entirely in this resistanceless core, and so the magnetic flux density it produces at the surface of the core will be $H_c a/r$. Since this is greater than H_c, the superconducting core will shrink to a smaller radius and this process will continue until the superconducting core contracts to zero radius, i.e. the whole wire is normal. However, it is not possible for the wire to become completely normal at a current i_c because if the wire were normal throughout, the current would be uniformly distributed over the cross section and the magnetic field strength inside the wire at a distance r from the centre would be less than the critical field, so the material could not be normal.

It therefore appears that, at the critical current, the wire can be neither wholly superconducting nor wholly normal, and that a state in which a normal sheath surrounds a superconducting core is not stable. In fact, at the critical current, the wire goes into an intermediate state of alternate superconducting and normal regions each of which occupies the full cross-section of the wire.[†] The current passing along the wire now has to flow through the normal regions, so at the critical current the resistance should jump from zero to some fraction of the resistance of the completely normal wire. Experiments show that a considerable resistance does indeed suddenly appear when the current is raised to the critical value (Fig. 7.4), the resistance jumping to between 0·6 and 0·8 of the full normal resistance. The exact value depends on factors such as the temperature and purity of the wire.

The detailed shapes of the normal and superconducting regions which appear when a current exceeding the critical current is passed along a

[†] F. London, *Superfluids*, vol. 1, Dover Publications Inc., New York, 1961.

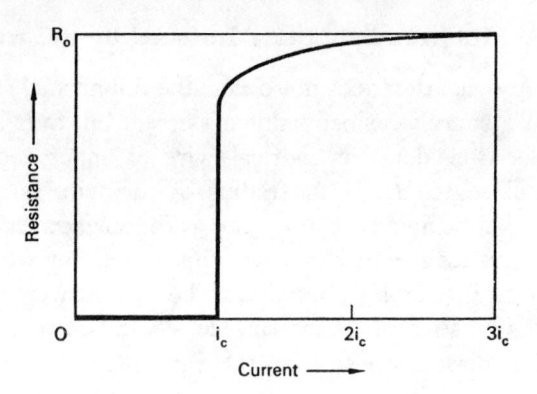

FIG. 7.4. Restoration of resistance to a wire by a current.

wire have not yet been determined experimentally, and it is a complicated problem to deduce them from theoretical considerations. The configuration shown in Fig. 7.5a is one which has been recently proposed on a theoretical basis, and for which there is some supporting experimental evidence.

It can be seen from Fig. 7.4 that as the current is increased above the critical value i_c the resistance of the wire gradually increases and approaches the full normal resistance asymptotically. London suggested that when the current is increased above the critical value i_c the in-

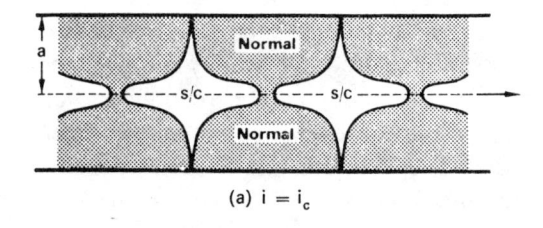

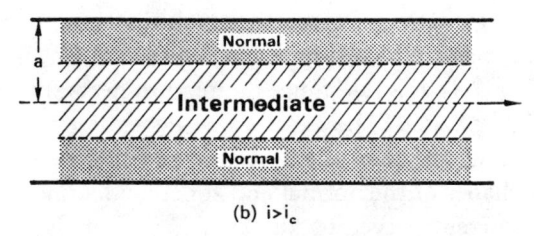

FIG. 7.5. Suggested cross-section of cylindrical wire carrying current in excess of its critical current (based on Baird and Mukherjee, and London).

termediate state contracts into a core surrounded by a sheath of normal material whose thickness increases as the current increases, so that the total current is shared between the fully resistive sheath and the partially resistive intermediate core (Fig. 7.5b). This model predicts a resistance increasing smoothly with the current in excess of i_c.

The sudden appearance of resistance, either by thermal propagation or by the appearance of an intermediate state when the critical current is exceeded, can make the measurement of the critical current of a conducting wire a rather precarious experiment. As the measuring current through the specimen is increased, a resistance R suddenly appears when the critical value i_c is exceeded. Power $i_c^2 R$ is then generated in the specimen, and if i_c is large and R is not very small, the heating may be enough to melt the wire, unless the current is reduced very quickly. In fact, superconducting wires can act like very efficient fuses with a sharply defined burn-out current.

CHAPTER 8

THE SUPERCONDUCTING PROPERTIES OF SMALL SPECIMENS

IT WAS pointed out in Chapter 2 that the penetration depth λ is very small, and that most superconducting specimens have dimensions which are very much greater than λ. Sometimes, however, a situation arises, as for example with thin films or fine wires, in which one or more of the dimensions of the specimen is comparable with λ. We shall see in this chapter that the superconducting properties of such specimens are in some ways significantly different from those of large specimens.

8.1. The Effect of Penetration on the Critical Magnetic Field

We saw in Chapter 4 that if a specimen of superconducting metal is in a uniform applied magnetic field H_a, the superconductor is driven into the normal state when H_a is increased above a critical value H_c. From a thermodynamic point of view, this is because the Gibbs free energy of a superconducting specimen is changed by an amount $-\int_0^{H_a} \mu_0 M dH_a$ in an applied field H_a, where M is the induced magnetic moment.

In the superconducting state M is negative, so the free energy is increased, and if this increase is sufficient to make the free energy in the superconducting state exceed that in the normal state, the specimen becomes normal. The magnetic moment M is equal to $\int I dV$, where V is the volume of the specimen and I is the intensity of magnetization given by

$$B = \mu_0 H + \mu_0 I. \tag{8.1}$$

In Chapter 4 it is assumed that $B = 0$ everywhere inside the superconductor, so that $I = -H$ and $M = -HV$; in other words, it was assumed that the magnetic moment per unit volume is independent of the shape and size of the specimen. It follows from this that the critical magnetic field is given by

$$\tfrac{1}{2}\mu_0 H_c^2 = g_n - g_s \quad [\text{see (4.3)}], \tag{8.2}$$

where g_n and g_s are the free energies per unit volume of the normal and superconducting phases in zero magnetic field. H_c should therefore be independent of the size of the specimen.

This argument needs modifying when penetration of the field is taken into account. We saw in Chapter 2 that B does not fall abruptly to zero at the surface of the specimen, but decreases with distance (x) into the interior of the specimen approximately as $e^{-x/\lambda}$, where λ, the "penetration depth", is about 5×10^{-6} cm in most superconductors. It follows that just inside the surface B is not zero, and that in this region the value of I given by (8.1) is no longer equal to $-H$. Instead, the magnitude of I increases from zero at the surface to the value H in the interior of the specimen, and as a result the magnitude of the magnetic moment M is less than it would be if λ were zero. Hence, for a given value of H_a, the magnetic contribution to the free energy, $-\int_0^{H_a} \mu_0 M dH_a$ or $-\tfrac{1}{2}\mu_0 M H_a$, is less than it would be if penetration did not occur, and the applied magnetic field has to be increased beyond the value given by (8.2) before the transition to the normal state can take place. In other words, the critical magnetic field is increased as a result of the penetration of the magnetic flux. The magnitude of the increase depends on the reduction in the magnetic moment M, which in turn depends on the dimensions of the specimen relative to the penetration depth λ. The effect is only noticeable if the volume contained within a distance λ of the surface is comparable with the total volume of the specimen.

8.2. The Critical Field of a Parallel-sided Plate

The effect of penetration on the critical magnetic field of a specimen is most easily illustrated by reference to the case of a parallel-sided plate with a magnetic field H_a applied parallel to the surfaces of the plate. The length of the plate in the direction of the field and its width are both supposed much larger than its thickness (Fig. 8.1). This particular geometry is chosen partly for its practical importance and partly because the flux distribution within the plate can be easily calculated with the aid of the London equations.

If the direction normal to the surfaces of the plate is chosen as the

x-direction, the variation of the flux density with x as given by the London equations (§ 3.2.1) is

$$B(x) = \frac{\cosh(x/\lambda_L)}{\cosh(a/\lambda_L)} \mu_0 H_a, \qquad (8.3)$$

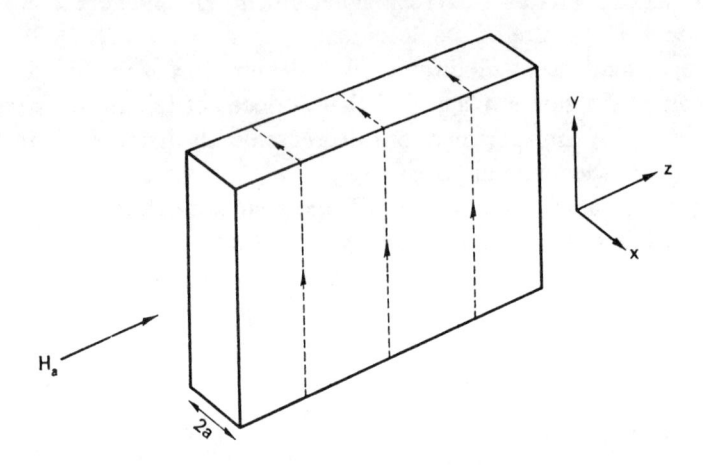

FIG. 8.1. Superconducting plate of thickness $2a$ with a magnetic field parallel to its surfaces. Its length and height are assumed much greater than $2a$. The broken lines show the direction of the screening currents.

where x is measured from the mid-plane of the film and the thickness of the film in $2a$. We saw in § 3.2 that, for large specimens, the variation of B with x predicted by the London equations is such that λ_L satisfies the general definition of λ given by (2.2). We shall therefore neglect the distinction between λ_L and λ in what follows, and write the solutions of the London equations in terms of λ. The dependence of B on x is illustrated graphically in Fig. 8.2, for the case where the thickness of the film is much greater than λ. Since the value of the magnetic field strength is H_a throughout the film,† the intensity of magnetization is equal to $(B/\mu_0) - H_a$, and the magnetic moment M (given by $\int I dV$) and the magnetic contribution to the free energy (given by $-\int_0^{H_a} \mu_0 M dH_a$ or $-\tfrac{1}{2}\mu_0 M H_a$) are both proportional to the cross-hatched area. For the case of Fig. 8.2 this is only marginally different from the value it would have if λ were zero. However, if $a \sim \lambda$, the situation is as shown in Fig. 8.3, and it is clear

† See Appendix A.

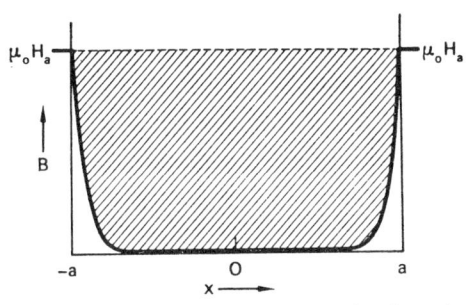

FIG. 8.2. Variation of B with distance normal to the surface for a plate of thickness $2a$ ($2a \gg \lambda$). The cross-hatched area is proportional to the magnetic moment and to the magnetic free energy.

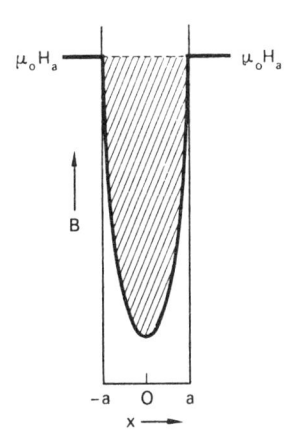

FIG. 8.3. Variation of B with distance normal to the surface for a plate of thickness $2a$ ($a \sim \lambda$).

that the reduction in the cross-hatched area relative to the value it would have if λ were zero is now very considerable. At every point the magnetization is $I(x) = \dfrac{B(x)}{\mu_0} - H_a$, so the value of M, the magnetic moment per unit area of the plate, is given by

$$\begin{aligned} M &= \int_{-a}^{a} \left[\frac{B(x)}{\mu_0} - H_a \right] dx \\ &= 2H_a \int_{0}^{a} \left[\frac{\cosh(x/\lambda)}{\cosh(a/\lambda)} - 1 \right] dx \\ &= -2aH_a \left[1 - \frac{\lambda}{a} \tanh \frac{a}{\lambda} \right] \end{aligned} \qquad (8.3\text{a})$$

It is convenient to write $M = -2akH_a$ so that

$$k = \left[1 - \frac{\lambda}{a}\tanh\frac{a}{\lambda}\right] \tag{8.4}$$

Due to the penetration of the flux, the effective susceptibility of the plate is $-k$ instead of -1. Note that k is positive, and that $k = 1$ if $\lambda = 0$. The magnetic contribution to the Gibbs free energy per unit area of the plate is

$$-\tfrac{1}{2}\mu_0 M H_a = \mu_0 a k H_a^2$$

so that the critical magnetic field is given by

$$\mu_0 a k H_c^2 = G_n - G_s = 2a(g_n - g_s)$$

where G_n and G_s refer to unit area of the plate, and g_n and g_s to unit volume as before. If we use H_c' to denote the critical field of a plate of thickness $2a$ for a penetration depth λ and H_c the critical field if λ were zero, then H_c' is given by

$$\tfrac{1}{2}\mu_0 k H_c'^2 = g_n - g_s \tag{8.5}$$

compared with $\tfrac{1}{2}\mu_0 H_c^2 = g_n - g_s$ given by (8.2).

Hence

$$H_c' = k^{-\frac{1}{2}} H_c = H_c\left[1 - \frac{\lambda}{a}\tanh\frac{a}{\lambda}\right]^{-\frac{1}{2}} \tag{8.6}$$

i.e. the critical field is increased due to the penetration of the flux. We have referred to H_c as the critical magnetic field for the case of zero penetration, but since (8.6) involves only the ratio a/λ, we could equally well regard H_c as the critical magnetic field for an infinitely large specimen. For this reason H_c is usually referred to as "the bulk critical field".

Equation (8.6) can be simplified for the cases of $a \gg \lambda$ or $a \ll \lambda$. If $a \gg \lambda$,

$$1 - \frac{\lambda}{a}\tanh\frac{a}{\lambda} \simeq 1 - \frac{\lambda}{a}$$

and
$$H_c' \simeq H_c\left(1 - \frac{\lambda}{a}\right)^{-\frac{1}{2}} \simeq H_c\left(1 + \frac{\lambda}{2a}\right), \tag{8.7}$$

which has the following simple physical interpretation. If $a \gg \lambda$ (8.4) takes the form $k \simeq 1 - (\lambda/a)$, so that $M \simeq -2(a - \lambda)H_a$. This is as if the intensity of magnetization I had remained equal to $-H_a$ throughout the plate, but the thickness of the plate had shrunk to $2(a - \lambda)$, i.e. as if each surface of the plate had receded inwards a distance λ. This is in accordance with the phenomenological definition of λ given by (2.2). For the other extreme case of a $a \ll \lambda$,

$$1 - \frac{\lambda}{a} \tanh \frac{a}{\lambda} \simeq \frac{a^2}{3\lambda^2}$$

so that
$$H'_c \simeq \sqrt{(3)} \frac{\lambda}{a} H_c. \tag{8.8}$$

It is important to get some idea of the order of magnitude of the increase in the critical field due to penetration of the flux. We have seen (§ 2.4.1) that the penetration depth obeys the relationship

$$\lambda = \lambda_0 \left\{ 1 - \left(\frac{T}{T_c}\right)^4 \right\}^{-\frac{1}{2}},$$

where λ_0 is about 500 Å (i.e. 500×10^{-10} m) for most pure metals. Equation (8.7) now shows that, for temperatures not too near the transition temperature, the increase in H_c will only be significant (say 10% or more) if the thickness of the plate is about 5000 Å or less, i.e. if we are dealing with a thin film. For the case of a *very* thin film, of thickness 100 Å or so (8.8) shows that H'_c may exceed H_c by an order of magnitude, especially if the temperature is close to the critical temperature.

8.3. More Complicated Geometries

Although the basic physics is the same as for the rectangular plate, the effect of penetration on the magnetic properties of a cylinder or sphere is much more difficult to calculate, the case of the cylinder involving Bessel functions. When the dimensions of the specimen are much greater than λ, we may, however, make use of the simple argument discussed immediately after (8.7), according to which the effect of penetration is as if the magnetization remained equal to $-H_a$ throughout the specimen but the surface of the cylinder were to recede inwards by a distance λ. Using this argument it may easily be shown that the critical magnetic field of a long cylinder of a radius a with the field applied parallel to its axis is given by

$$H'_c = H_c \left(1 + \frac{\lambda}{a}\right).$$

For the case of $a \ll \lambda$, it can be shown† that $H'_c = \sqrt{(8)}(\lambda/a)H_c$.

† For example, D. Shoenberg, *Superconductivity*, C.U.P., 1962, p. 234.

8.4. Limitations of the London Theory

It is clear from (8.6) to (8.8) that, since λ is of the order of 10^{-5} cm, the effect of penetration will be too small to produce any appreciable effect on the critical magnetic field unless one of the dimensions of the sample at right angles to the field is about 5000 Å or less, as occurs in the case of a thin film.

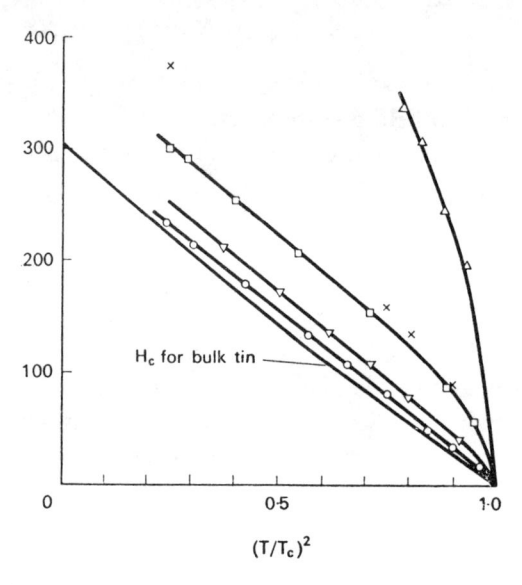

FIG. 8.4. Variation of parallel critical field with reduced temperature for tin films of various thicknesses ($\triangle$, 1000 Å; $\square$, 2000 Å; $\triangledown$, 5000 Å; $\bigcirc$, 10000 Å). Also shown are points calculated from an effective penetration depth as prescribed by Ittner ($\times$, 2000 Å). (After Rhoderick.)

The dependence of the critical magnetic field of thin films on their thickness is quite pronounced, as can be seen from Fig. 8.4. The critical fields of these films were determined by observing the restoration of resistance by a magnetic field parallel to the surface of the film. In the case of a film 1000 Å thick, the critical magnetic field close to the critical temperature is over an order of magnitude greater than that for the bulk metal. The increase relative to the bulk field is most marked near the critical temperature because, as was pointed out in Chapter 2, the penetration depth is found experimentally to vary with temperature approximately as $[1 - (T/T_c)^4]^{-\frac{1}{2}}$ and becomes infinite as T approaches T_c.

A comparison of the experimental results with the predictions of the London theory as expressed by (8.6) shows good qualitative agreement, but quantitative comparison is not easy. One could, of course, assume the truth of (8.6) and use this equation to calculate an effective value of λ from the measured values of H_c, H_c' and a. The important point, however, is whether the value of λ obtained in this way agrees with the values obtained from the magnetization of bulk specimens and shows the same temperature variation. Only in this way is it possible to check the validity of the London theory. It should be remembered that the measurements of λ obtained from the magnetization of bulk specimens, discussed in Chapter 2, are independent of any particular penetration law and depend only on the definition of λ given by

$$\lambda = \frac{1}{B(0)} \int_0^\infty B(x)\, dx,$$

where x is measured from the surface. As we pointed out on page 97, (8.7) is consistent with this definition.

Equation (8.6), however, is derived from (8.3), which is the penetration law predicted by the London theory. Whether or not (8.6) correctly predicts the critical fields for films of *any* thickness, using the value of λ determined from bulk measurements is therefore equivalent to verifying the truth of the penetration law expressed by (8.3).

Attempts to correlate theory and experiment in this way have been only partially successful. Figure 8.4 shows points calculated for a film of thickness 2000 Å using for λ_0 the value 520 Å obtained by magnetization measurements of bulk specimens and assuming a variation of λ with temperature given by

$$\lambda(T) = \lambda_0 \left[1 - \left(\frac{T}{T_c}\right)^4\right]^{-\frac{1}{2}}.$$

It is seen that the theoretical points lie consistently above the experimental ones.

To account for the discrepancy we must recall how the London theory originated. It is essentially a *phenomenological* theory, that is to say, it was introduced because it gives a fairly good description of the Meissner effect. It is important that we should not accord the London theory the status of, say, Maxwell's equations, which are believed to be exact expressions of inviolable physical laws. One obvious limitation is that it is essentially a classic:l theory which treats the electrons as classical particles, although we should expect quantum effects to be significant. Two important assumptions made in the London theory are that the penetration depth λ_L is independent of the strength of the applied magnetic field and also of the dimensions of the specimen. It should not really surprise us if the first of these assumptions turns out not to be strictly true. As can be seen from (3.13), it is equivalent to assuming that the effective number of superconducting electrons is independent of the applied field. However, the application of a magnetic field is known to modify the behaviour of electrons in a profound way, so we might expect it to change the degree of order, i.e. the effective number of superelectrons. It has been shown experimentally by Pippard that the penetration depth does in fact increase with applied magnetic field, although in bulk superconductors the effect is not large except near the transition temperature. The London theory is therefore essentially a *weak field* theory. The effect of such a dependence of λ on the magnetic field is that the magnetic moment M of a film is no longer linearly proportional to H_a, as (8.3a) would predict if λ were constant, but increases less rapidly. The graph of M against H_a is therefore non-linear, as shown in Fig. 8.5 and

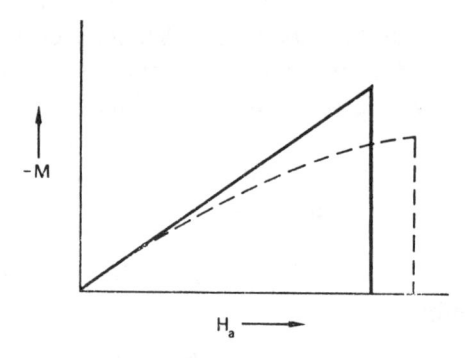

Fig. 8.5. Magnetization curve of a thin superconductor. ——— assuming λ independent of H_a. ————— assuming λ increases with H_a.

H_a has to be increased further before the free energy of the superconducting state becomes equal to that of the normal state; consequently H_c is increased.

The second assumption mentioned above, that the penetration depth is independent of the dimensions of the specimen, seems less open to objection, and it is difficult to explain in simple terms why it should not be correct. Suffice it to say that, as was seen in Chapter 6, superconducting electrons do not behave completely independently of each other, but exhibit "long range order" extending for a distance known as the coherence length ξ, which in a bulk superconductor is about 10^{-4} cm. If the dimensions of the specimen are less than this bulk coherence range, the value of ξ is reduced, and various properties of the superconductor (among them the penetration depth) are modified.

Various attempts have been made, by Ittner, Tinkham and others, to "patch up" the London theory in so far as its predictions regarding the critical fields of thin films are concerned by incorporating in it the field-dependence, temperature-dependence and size-dependence of the penetration depth which result from more recent theories, such as the microscopic theory of Bardeen, Cooper and Schrieffer (see Chapter 9). The success of these attempts is only moderate, and since they incorporate into the London theory new features which are in fact foreign to it, the procedure is not entirely satisfactory. It seems far preferable to look for a theory, such as the Ginzburg–Landau theory, in which the necessary non-linearity of M with respect to H and size-dependence of λ are inherent.

8.5. The Ginzburg–Landau Theory

The Ginzburg–Landau theory[†] is an alternative to the London theory. To a certain extent it is a phenomenological theory also, in the sense that it makes certain *ad hoc* assumptions whose justification is that they correctly describe the phase transition in zero field, but unlike the London theory, which is purely classical, it uses quantum mechanics to predict the effect of a magnetic field. The Ginzburg–Landau theory involves a good deal of algebra and a complete description of it would take us far beyond the scope of this book. However, we will try to give a brief description of what it is about, and of some of its more important predictions.[‡]

The first assumption of the Ginzburg–Landau theory is that the behaviour of the superconducting electrons may be described by an "effective wave function" Ψ, which has the significance that $|\Psi|^2$ is equal to the density of superconducting electrons. It is then assumed that the free energy of the superconducting state differs from that of the normal state by an amount which can be written as a power series in $|\Psi|^2$. Near the critical temperature it is sufficient to retain only the first two terms in this expansion. Ginzburg and Landau then point out that if, for any reason, the wave function Ψ is not constant in space but has a gradient, this gives rise to kinetic energy whose origin is the same as that of the

[†] V. L. Ginzburg and L. D. Landau, *J.E.T.P.* **20**, 1064 (1950).

[‡] An introduction to the Ginzburg–Landau theory is given by A. D. C. Grassie, *The Superconducting State*, Sussex University Press, 1975.

kinetic energy term $(\hbar^2/2m)\nabla^2\Psi$ which appears in Schrödinger's equation for a particle of mass m. To take account of this, an additional term, proportional to the square of the gradient of Ψ, is added to the expression for the free energy of the superconducting phase. The effect of a magnetic field is introduced by resorting to a theorem in classical mechanics which states that the effect of the Lorentz force ($q\mathbf{v} \times \mathbf{B}$) on the motion of a charged particle in a magnetic field $\mathbf{B}$ may be completely accounted for† by replacing the momentum $\mathbf{p}$, wherever it occurs in the expression for the kinetic energy, by the more complicated expression $\mathbf{p} - q\mathbf{A}$. Here $\mathbf{A}$ is the magnetic vector potential defined by $\mathbf{B} = \text{curl } \mathbf{A}$. To make the transition to quantum mechanics, $\mathbf{p}$ is replaced by the operator $-i\hbar$ grad. The total magnetic contribution to the free energy of the superconducting state is therefore given by

$$\frac{\hbar^2}{2m} \int \Psi^* \left[i \text{ grad} + \frac{eA}{\hbar} \right]^2 \Psi \, dV,$$

the integral being taken over the whole volume V of the specimen.

The central problem of the Ginzburg–Landau approach is now to find functions $\Psi(x, y, z)$ and $\mathbf{A}(x, y, z)$ which make the total free energy of the specimen a minimum subject to appropriate boundary conditions. For weak magnetic fields the problem is easily soluble and reduces to the same form as the London equations. In a strong magnetic field the equations are only soluble by numerical means. In the case of an infinitely thick plate with the applied magnetic field parallel to the surface, the solution predicts that $|\Psi|^2$ is constant in the interior of the plate but falls off towards the surface by an amount which increases with the applied

---

† This can be seen by the following simple argument. Suppose that a particle of charge q is moving in a field-free region with velocity $\mathbf{v}_1$ and that a magnetic field is applied at time $t = 0$. The field can only build up at a finite rate, and while it is changing there will be an induced electric field which satisfies Maxwell's relation curl $\mathbf{E} = -\dot{\mathbf{B}}$. If $\mathbf{A}$ is the vector potential, curl $\mathbf{E} = -(d/dt)$ (curl $\mathbf{A}$), and integration with respect to the space coordinates gives $\mathbf{E} = -(d\mathbf{A}/dt)$ apart from a constant of integration which does not concern us. Hence the momentum at time t is given by

$$m\mathbf{v}_2 = m\mathbf{v}_1 + \int_0^t q\mathbf{E} \, dt = m\mathbf{v}_1 - q \int_0^t \frac{d\mathbf{A}}{dt} \, dt$$

$$= m\mathbf{v}_1 - q\mathbf{A},$$

so that $m\mathbf{v}_2 + q\mathbf{A} = m\mathbf{v}_1$. Hence the vector $\mathbf{p} = m\mathbf{v} + q\mathbf{A}$ is conserved during the application of the field and must be regarded as the effective momentum when a magnetic field is present. However, the kinetic energy ε depends only on $m\mathbf{v}$, and if $\varepsilon = f(m\mathbf{v})$ before the field is applied, we must write $\varepsilon = f(\mathbf{p} - q\mathbf{A})$ in the presence of the field.

magnetic field. Since the penetration depth depends on the number of superconducting electrons at the surface, as in the London theory, and hence on $|\Psi|^2$, we at once get a field-dependent penetration depth. In the case of a thin film, because of the boundary conditions, the variation of Ψ with x depends on the thickness of the film, and since λ again depends on $|\Psi|^2$ the penetration depth is a function of the thickness of the film. Thus the two missing elements in the London theory are automatically provided by Ginzburg and Landau. The critical magnetic field can be calculated by the usual method of equating the free energy of the film in the superconducting state with that in the normal state. The general expression for H_c' is complicated, but simplifies in two special cases:

(i) $a \gg \lambda$

In this case

$$H_c' = H_c\left(1 + \frac{\alpha\lambda}{2a}\right),$$

where $2a$ is the thickness of the film, λ is the penetration depth in a *weak* magnetic field, and α is a coefficient very close to unity. This is essentially the same as the London result.

(ii) $a < \lambda$

In this case Ψ is approximately constant throughout the film, and $H_c' = \left[\frac{\sqrt{(6)}\lambda}{a}\right]H_c$. This differs from the London result (8.8) by a factor of $\sqrt{2}$. A surprising consequence of the Ginzburg–Landau theory is that in this case Ψ falls *gradually* to zero as H approaches H_c' so that the transition is a second-order one. (It should be remembered that in bulk super-conductors the transition is second order in the absence of a magnetic field but first order in the presence of a field.) The transition is first- or second-order according as $a \gtrless \frac{\sqrt{(5)}\lambda}{2}$. This prediction of the theory has been amply confirmed by Douglass using the tunnelling technique (see Chapter 10).

The Ginzburg–Landau theory predicts magnitudes of the critical magnetic fields which are not very different from those of the London theory in the limiting cases of very thick and very thin films. In the intermediate case, where solutions have to be obtained numerically, the

Ginzburg–Landau theory does not give a significantly better fit with experiment than the London theory if values of λ obtained from measurements on large specimens are used. The great success of the Ginzburg–Landau theory is that it correctly predicts the change from first- to second-order transitions with decreasing thickness, which the London theory does not.

Various refinements have been incorporated into the Ginzburg–Landau theory to improve its quantitative predictions. The general situation with respect to the theoretical interpretation of critical fields of thin films is still, however, not really satisfactory.

8.6. Edge Effects

One important consequence of the dependence of the critical magnetic field of a film on its thickness is that the sharpness of the magnetic transition for a thin film depends very much on the nature of its edges. As a rule films are prepared in the form of a strip by evaporating the superconducting metal on to an insulating base (or "substrate") through a mask which behaves like a stencil. Because the mask is never in precise contact with the substrate, and also because the metal atoms are able to wander about on the substrate before they finally come to rest, the edges of the film are never completely sharp but always tend to be tapered, as shown in Fig. 8.6. The tapered edges, being thinner than the rest of the

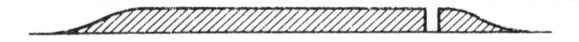

FIG. 8.6. Typical cross-section of evaporated film, showing tapered edges. The right-hand edge has been trimmed to produce a well defined rectangular geometry.

film, have a higher critical magnetic field and if the film is tested for superconductivity by passing a current through it and seeing whether any voltage difference appears across it ends, the edges will remain superconducting and give rise to zero resistance even when the rest of the film is normal. This has two consequences. First, the magnetic field strength at which a voltage difference appears may be considerably greater than the true critical magnetic field of the film; and second, because the edges are unlikely to be perfectly uniform along the length of the strip, the transition from zero to full normal resistance may extend over a considerable range of values of the magnetic field. To obtain sharp resistance transitions at the true value of magnetic field, the edges of a film are usually trimmed as shown on the right-hand side of Fig. 8.6. The effect of trimming a film is shown in Fig. 8.7.

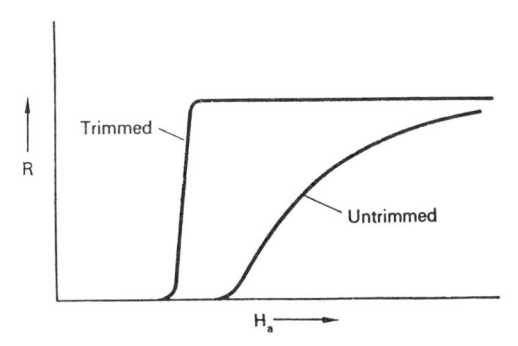

FIG. 8.7. Effect of trimming the edges on resistance transition of thin film.

8.7. Transitions in Perpendicular Magnetic Fields

So far we have limited the discussion to the case where the magnetic field is applied parallel to the surface of the film. Since films are always thin compared with their other two dimensions, a film is essentially a "long, thin specimen" in the sense used in Chapter 6, and demagnetizing effects are quite negligible. However, if a magnetic field is applied perpendicular to the surface of the film, demagnetizing effects become very important. If the film takes the form of a relatively narrow strip whose width w is much less than its length but much greater than its thickness d, then we may approximate the strip by an elliptical cylinder whose cross-section has axes w and d. The demagnetizing coefficient for such a geometry is given by $n \simeq 1 - (d/w)$. According to Chapter 6, we expect the film to enter the intermediate state if exposed to a perpendicular field of magnitude $H_c(1 - n) \simeq H_c(d/w)$. Typical values encountered in practice are $w \sim 10^{-1}$ cm and $d \simeq 10^{-5}$ cm, so that $(d/w) \sim 10^{-4}$. We should therefore expect the film to be driven normal by an extremely small perpendicular field—in fact, the earth's field should be more than sufficient. In practice this is not so, the reason being that the theory of the intermediate state outlined in Chapter 6 is no longer valid when the thickness of the film is comparable with the penetration depth, mainly because the concept of surface energy discussed in Chapter 6 now requires considerable modification. A complete discussion of the subject would take us far beyond the scope of this book. According to Tinkham, the intermediate state in thin films resembles the mixed state in type-II superconductors (see Chapter 12) and can be described in terms of current vortices.

Experimentally it is found that the transition to the normal state of thin films in perpendicular magnetic fields does occur at a significantly lower field strength than in the case of parallel fields, but not as low as the previous paragraph would indicate. For this reason, in carrying out studies of transitions in parallel magnetic fields, great care has to be taken to ensure that the film is accurately parallel to the applied field so that there is no component of the field perpendicular to the surface. In the case of resistive transitions this is easily accomplished by setting the magnetic field so that about half of the normal resistance is restored, and then rotating the magnet (or the specimen) until the resistance is a minimum, which means that the critical magnetic field has its maximum value.

8.8. Critical Currents of Thin Specimens

For large specimens we have already seen that the critical current can be calculated from the critical magnetic field by making use of Silsbee's criterion. This states that in the absence of an applied magnetic field the critical current is that current which produces at the surface of the specimen a magnetic field equal to the critical field H_c. Unfortunately this simple rule does not hold for specimens which have one or more dimensions comparable with λ, the most obvious reason being that even if some sort of modified Silsbee's rule were to hold we should not know whether to insert the bulk critical field H_c or the actual critical field for the small specimen H_c'. The problem is a complicated one and depends, among other things, on the current distribution in the film. We therefore need some way of calculating the current distribution, and for simplicity we shall follow the London theory, notwithstanding the limitations of this theory which we have already mentioned. An entirely adequate theory of critical currents in thin specimens has yet to be formulated, but the predictions of the London theory are sufficiently correct to give some qualitative indication of the effects that are to be expected.

A further difficulty is that we cannot adopt the simple thermodynamic approach of equating the free energies of the superconducting and normal phases because, in the presence of a transport current supplied by an external source, the transition is irreversible due to the fact that energy is continuously dissipated in the normal state. Fortunately a way round this difficulty was found by H. London, who pointed out that the existence of a critical magnetic field H_c for a bulk superconductor can be regarded as resulting from the existence of a critical current density $\mathcal{J}_c$, as we saw in Chapter 4. To illustrate this principle, consider the special case of a superconducting plate of thickness $2a$ with a magnetic field applied parallel to its surfaces (see Fig. 8.1). Suppose that the thickness of the plate lies along the x-direction, that the applied field is in the z-direction, and that the dimension of the plate in the y-direction is much greater than $2a$. Then to exclude flux from the interior of the plate the shielding currents must flow parallel to the y-axis, as shown in the figure. As we have already seen, the solution of the London equations gives

$$B(x) = \frac{\cosh(x/\lambda)}{\cosh(a/\lambda)} \mu_0 H_a,$$

where x is measured from the midplane of the plate. The current density $\mathcal{J}_y$ can be found from Maxwell's equation† curl $\mathbf{B} = \mu_0 \mathbf{J}$, which for the geometry shown in Fig. 8.1 simplifies to

$$\mathcal{J}_y = -\frac{1}{\mu_0}\frac{\partial B}{\partial x} = -\frac{H_a}{\lambda}\frac{\sinh(x/\lambda)}{\cosh(a/\lambda)};$$

B and $\mathcal{J}$ are plotted as functions of x in Fig. 8.8.

Notice that although B has the same direction throughout the plate, the current density has opposite directions in the two halves. The current density is greatest at the surfaces of the plate, where it has the magnitude $(H_a/\lambda)\tanh(a/\lambda)$. According to London's postulate, the plate is driven into the normal state when the current density at the surface reaches a critical value $\mathcal{J}_c$. However, we know that for thick specimens ($a \gg \lambda$) the plate is driven normal when $H_a = H_c$, and in this case $\tanh(a/\lambda) \to 1$, so the relationship between the critical current density $\mathcal{J}_c$ and the critical magnetic field is

$$\mathcal{J}_c = \frac{H_c}{\lambda}. \tag{8.9}$$

Now suppose that instead of applying an external field H_a to the plate we pass in the y-direction a current which has the magnitude $\mathcal{J}$ per unit width of the plate in the z-direction. Associated with this transport current there will be a flux density B both inside and outside the plate. Clearly, from the symmetry of the situation, the current density

---

† Note that we have written this equation in terms of **B** rather than the usual **H**. This is because **J** is not the "free" current density (see Appendix A) but the induced magnetization current.

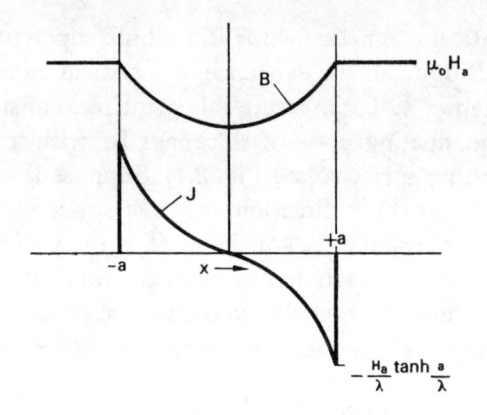

FIG. 8.8. Variation of B and $\mathcal{J}$ with x for plate of thickness $2a$ in uniform applied field H_a.

associated with the transport current must have the same direction in each half of the plate, but B will have opposite directions. We therefore need a solution of the London equations which has the form

$$B(x) = -B(-x) \quad \text{and} \quad \mathcal{J}(x) = \mathcal{J}(-x).$$

Such a solution is

$$\mathcal{J}(x) = \mathcal{J}(a) \frac{\cosh(x/\lambda)}{\cosh(a/\lambda)}$$

and

$$B = -\mu_0 \lambda^2 \operatorname{curl} \mathbf{J} = -\mu_0 \lambda^2 \frac{\partial \mathcal{J}_y}{\partial x}$$

$$= -\mu_0 \lambda \mathcal{J}(a) \frac{\sinh(x/\lambda)}{\cosh(a/\lambda)}; \tag{8.10}$$

B and $\mathcal{J}$ are plotted as functions of x in Fig. 8.9. $\mathcal{J}(a)$ is related to the current per unit width $\mathcal{I}$ by

$$\mathcal{I} = \int_{-a}^{a} \mathcal{J}(x)\,dx = 2\lambda\,\mathcal{J}(a) \tanh(a/\lambda)$$

The current density is again a maximum at the surface and if we assume that the plate will begin to go normal when $\mathcal{J}(a)$ has the value $\mathcal{J}_c = H_c/\lambda$, as before, then the critical current per unit width of the plate $\mathcal{I}'_c$, is given by

$$\mathcal{I}'_c = 2\lambda \mathcal{J}_c \tanh(a/\lambda)$$
$$= 2H_c \tanh(a/\lambda). \tag{8.11}$$

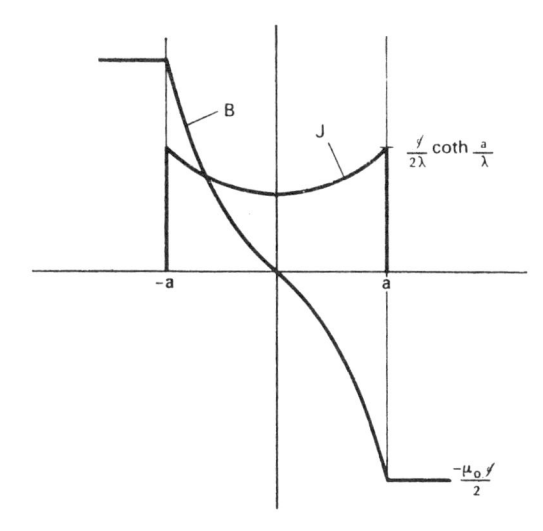

FIG. 8.9. Variation of B and J with x for plate of thickness $2a$ carrying current $\mathscr{I}$ per unit width.

If $a \gg \lambda$, $\tanh(a/\lambda) \to 1$, and the critical current per unit width becomes $\mathscr{I}_c = 2H_c$. In this case, from (8.10) the magnitude of the flux density at the surface is $\mu_0 H_c$, i.e. the magnetic field strength is H_c, which is in accordance with Silsbee's rule. In other words, *for plates which are much thicker than λ, regarding the destruction of superconductivity as determined by a critical current density J_c is in all respects equivalent to thinking of it as associated with a critical magnetic field H_c*. Note that the critical current per unit width of the plate is independent of its thickness, which is to be expected because all the current is concentrated within a penetration depth of the two surfaces.

In the case when a is comparable with λ, the critical current per unit width $\mathscr{I}'_c$ is, according to (8.11), given by

$$\mathscr{I}'_c = 2H_c \tanh(a/\lambda) = \mathscr{I}_c \tanh(a/\lambda), \qquad (8.12)$$

so that the critical current is *reduced* by the factor $\tanh(a/\lambda)$, in contrast with the fact that the critical magnetic field is *increased*. If $a \ll \lambda$, $\tanh(a/\lambda) \simeq (a/\lambda)$ and $\mathscr{I}'_c$ is proportional to the thickness of the film, as might be expected from the fact that in this case the current is almost uniformly distributed throughout the cross-section of the plate.

It is interesting to calculate the magnetic field at the surface of the film when superconductivity is destroyed by a current. From (8.10) the field strength is given by $|H(a)| = \lambda \mathcal{J}(a) \tanh(a/\lambda)$, and if $\mathcal{J}(a)$ is given by $\mathcal{J}(a) = \mathcal{J}_c = H_c/\lambda$, then

$$H(a) = H_c \tanh(a/\lambda).$$

Since $\tanh(a/\lambda) < 1$, we see that the magnetic field strength at the surface when superconductivity is quenched by a current, which we denote by H_l, is not only less than the magnitude of the externally applied field necessary to drive the film normal in the absence of a current H'_c, but is even less than the bulk critical field H_c, so that Silsbee's rule is not obeyed in any form. If $a \ll \lambda$, $H_l \simeq H_c a/\lambda$, and we have already seen from (8.8) that for this case $H'_c = \sqrt{(3)} H_c \lambda/a$, so that $H_l H'_c = \sqrt{(3)} H_c^2$. The Ginzburg–Landau theory gives essentially the same relationship between H_l, H'_c and H_c, but with a factor $4/3$ instead of $\sqrt{3}$.

8.9. Measurements of Critical Currents

Reliable measurements of the critical currents of thin films are hard to make for two reasons—first, because the attainment of a suitable geometry is difficult and, second, because it is not easy to prevent effects due to Joule heating confusing the issue.

The discussion of § 8.2 assumed that the film was infinitely wide in the z-direction, so that the current density was independent of z. For a film of finite width this is no longer true. If the thickness of the film were large compared with λ, the distribution of current over its surface would be exactly the same as occurs in an analogous situation in electrostatics, namely the distribution of charge on a conductor of the same geometry (see § 2.3). It is well known that charge tends to be concentrated where the radius of curvature is greatest, so it follows that in the case of a superconducting plate of rectangular geometry the current density will be greatest at the edges. If the thickness of the plate is comparable with λ, the analogy with electrostatics is no longer complete, but the current density is still greatest at the edges of the plate (or film). The critical current will therefore be very sensitive to the geometrical perfection of these edges and, as we saw in § 8.6, it is difficult to obtain perfectly sharp edges unless the films are trimmed.

A further complication which often arises when one tries to measure the critical currents of thin films is that if there happens to be a "weak spot", such a small region which is thinner or narrower than the rest of the film, or has slightly different metallurgical properties, then this weak spot will be driven normal before the true critical current of the film is reached, and the normal region may then spread by thermal propagation as discussed in § 7.2. This may be avoided by using current in the form of short pulses (say $1/10\,\mu$sec or even less) which do not allow time for any appreciable thermal propagation to take place, or by taking great care to see that the film is evaporated on to a substrate with a high thermal conductivity in good thermal contact with the helium bath.

The most careful experiments on critical currents of thin films to date seem to be those of Glover and Coffey, who calculated the current distribution from the London equations, and assumed the film to be driven normal when the current density at the edges of the strip reached $\mathcal{J}_c$. They found that their results were consistent with the existence of a critical current density and that the value of $\mathcal{J}_c$ for tin at 0°K was about 2×10^7 A cm^{-2}.

CHAPTER 9

THE MICROSCOPIC THEORY OF SUPERCONDUCTIVITY

SO FAR, we have taken a purely macroscopic view of superconductivity; that is to say, we have assumed that some of the electrons behave as superelectrons with the mysterious property that, unlike normal electrons, they can move through the metal without hindrance of any kind. We have then discussed what restrictions are placed on their collective behaviour by the laws of electromagnetism and thermodynamics. In this chapter we delve a little more deeply and take a microscopic view, trying to explain from first principles how it is that this property of the superelectrons arises. The complete microscopic theory of superconductivity is extremely complicated, and requires an advanced knowledge of quantum mechanics. All we can do within the scope of this book is to give a sketch of the physical principles involved.

9.1. Summary of the Properties of the Superconducting State

To obtain some clues as to the origin of superconductivity it is helpful to summarize the more important properties of superconductors.

9.1.1. Zero resistance

As we saw in Chapter 1, superconductors show zero resistance for direct currents and not-too-high frequency alternating currents. On the other hand, the reflection and absorption of visible radiation by a superconductor is the same in the superconducting as in the normal state (superconductors do not look any different when cooled below their critical temperature), and since optical properties can be related through Maxwell's equations to resistivity, we infer that at optical frequencies the superconducting state exhibits the same resistivity as the normal state.

The frequency at which resistance begins to appear corresponds to the microwave or long-wave infrared portions of the spectrum. The difference between the reflection coefficient of a superconductor in its superconducting and normal state is shown as a function of frequency for a few typical metals in Fig. 9.1. The curves are all roughly similar in shape, and show a steep fall to zero commencing at a frequency v_0 which varies from metal to metal. The value of v_0 for a particular metal depends on the temperature and tends to zero as the temperature rises towards the critical temperature, in a way reminiscent of the temperature-dependence of the critical magnetic field. Well below the critical temperature v_0 is substantially constant and has the value 3×10^{11} Hz for indium.

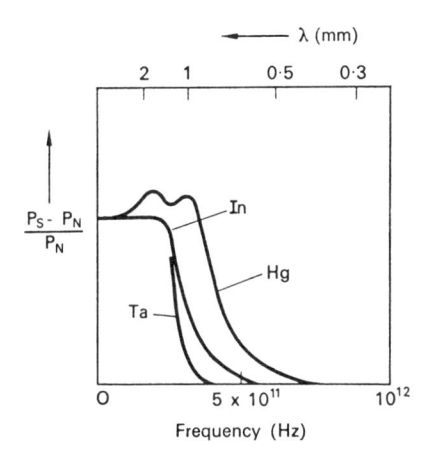

FIG. 9.1. Infrared reflection coefficient of various metals at 1·3°K (after Richards and Tinkham). The ordinate is proportional to the difference between the reflection coefficients in the superconducting and normal states.

9.1.2. Crystal structure

Studies of the crystal structure of superconductors by X-ray crystallography as the metal is cooled below the critical temperature have revealed no change in the lattice structure, either as regards the symmetry of the lattice or the actual lattice spacing. It has also been found that properties which depend on vibrations of the crystal lattice, such as the Debye temperature θ and the lattice contribution to the specific heat, are the same in the normal and superconducting phases. It is fairly clear, therefore, that superconductivity is not associated with any change in the properties of the crystal lattice.

9.1.3. Electronic specific heat

As we have seen in Chapter 5, the specific heat of a normal metal at low temperature has the form

$$C_n = A\left(\frac{T}{\theta}\right)^3 + \gamma T$$

where the first term refers to the specific heat of the lattice and the second to the conduction electrons. However, if a superconductor is cooled below its critical temperature in the absence of a magnetic field, two differences occur. First, there is a very significant jump in the value of the specific heat without the appearance of any latent heat, which shows that the transition is a second-order one; and, second, for very low temperatures the specific heat is given by an expression of the form

$$C_s = A\left(\frac{T}{\theta}\right)^3 + ae^{-b/kT}.$$

The value of A is unaltered when the superconductor is driven normal by a magnetic field, confirming that the lattice specific heat does not change, but the contribution from the conduction electrons is quite different. It is clear, then, that the superconducting state involves some quite drastic change in the behaviour of the conduction electrons.

9.1.4. Long-range order

There is considerable evidence from various sources that the superconducting electrons possess some sort of long-range order. For example, the existence of a positive surface energy mentioned in Chapter 6 leads to the conclusion that a boundary between normal and superconducting regions cannot be sharp, and that there is a gradual change in properties between them over a distance which in type-I superconductors is about 10^{-4} cm. In terms of the two-fluid model discussed in Chapter 1, we may say that the concentration of superelectrons cannot fall to zero abruptly at a boundary between superconducting and normal regions, but decreases gradually over a distance ξ, termed by Pippard the coherence length, which in pure metals is of the order of 10^{-4} cm. Figuratively speaking, one might say that the superelectrons are somehow aware of the existence of other electrons within a range of about 10^{-4} cm, and modify their behaviour accordingly. For this reason superconductivity is often referred to as a *co-operative* phenomenon. As will be seen in

Chapter 11, there are even more striking interference phenomena which show that superconducting electrons maintain some sort of phase coherence over macroscopically large distances, of the order of metres.

9.1.5. The isotope effect

Another experimental result which made a significant impact on the theory of superconductivity was the discovery in 1950 by Maxwell, and independently by Reynolds, Serin, Wright and Nesbitt, that, if measurements are carried out on specimens made from different isotopes of a given element, they are found to have different critical temperatures. In many cases the critical temperature turns out to be inversely proportional to the square root of the isotopic mass. Thus, although the atomic lattice does not itself show any change in properties between the normal and superconducting states, as discussed in § 9.1.2, it must nevertheless play a very important role in determining the change in behaviour of the conduction electrons.

9.1.6. The Meissner effect

In many ways the Meissner effect is the most fundamental property of superconductors; it incorporates the property of zero resistance inasmuch as the diamagnetic screening currents are constant in time and do not die away as long as the applied field remains unchanged. However, although an explanation has been given within the framework of the microscopic theory we are about to develop, it is difficult to interpret on the level of this book, and is not of much assistance for an elementary understanding of superconductivity.

9.2. The Concept of an Energy Gap

The reader who knows something about semiconductors may have noticed a similarity between Fig. 9.1 and the corresponding curve for the absorption of infrared radiation by a semiconductor. The broad features of the two curves are almost identical, but with the important distinction that in the case of a semiconductor the absorption "edge" (i.e. the sudden increase in absorption) occurs at a frequency which is about three orders of magnitude greater than in the case of a superconductor.

The explanation of the absorption of radiation by a semiconductor is now well established. In a semiconductor there is a separation in energy, or "energy gap", between the top of the (full) valence band of electron energy levels and the bottom of the (empty) conduction band. If the frequency of the incident radiation is sufficiently high for the photon energy $h\nu$ to exceed the energy gap, the photons will be able to excite an electron from the valence band to the conduction band and at the same time be absorbed in the process. It is natural to postulate that something similar occurs in the case of a superconductor, and that radiation is heavily absorbed when the photon energy is sufficient to excite electrons across an energy gap of some sort. Since in a superconductor absorption occurs for frequencies greater than 10^{11} Hz, the energy gap must be of the order of 10^{-4} eV. It may be observed that if we express this energy gap in the form kT, then T is about 1°K, which is of the order of magnitude of superconducting critical temperatures. The significance of this will appear later.

More evidence for the existence of some sort of energy gap in the electron levels comes from specific heat data. As we have already pointed out, at very low temperatures the contribution to the specific heat due to the conduction electrons in the superconducting state is proportional to $e^{-b/kT}$. This is precisely the form to be expected if there is a gap in the range of energies available to an electron. As the temperature is raised, electrons are thermally excited across the gap and for each of these electrons an amount of energy equal to the energy gap E_g is absorbed in the process. It follows from a simple application of statistical mechanics that at a temperature T the number of electrons in energy levels above the gap is proportional to $e^{-E_g/2kT}$, where k is the Boltzmann's constant, and the thermal energy absorbed in exciting these conduction electrons is therefore proportional to $E_g e^{-E_g/2kT}$. The specific heat associated with this process is proportional to the derivative of the energy with respect to temperature, i.e. to $(1/T^2)e^{-E_g/2kT}$. The T^{-2} term varies much more slowly with T than the exponential term, so the variation of specific heat with temperature should be very nearly exponential.

Additional evidence for the existence of an energy gap in the superconducting state is provided by the tunnelling experiments to be described in Chapter 10.

It should be stated, however, that under certain special circumstances a superconducting metal may not possess an energy gap. We bring up this point again in § 9.3.8. These "gapless" superconductors are not typical, and under normal circumstances all elemental superconductors and most alloys exhibit a well-defined energy gap.

9.3. The Bardeen–Cooper–Schrieffer Theory

9.3.1. Restatement of the problem

To recapitulate, any successful microscopic theory of superconductivity must be able to explain the following:

(i) Superconductivity is essentially bound up with some profound change in the behaviour of the conduction electrons which is marked by the appearance of long range order and a gap in their energy spectrum of the order of 10^{-4} eV.
(ii) The crystal lattice does not show any change of properties, but must nevertheless play a very important part in establishing superconductivity because the critical temperature depends on the atomic mass (the isotope effect).
(iii) The superconducting-to-normal transition is a phase change of the second order.

The long-range order noted in (i) clearly means that the electrons must interact with each other. It has, of course, been appreciated for a long time that the conduction electrons in a metal interact very strongly through their coulomb repulsion, and it is surprising that the ordinary free-electron theory of metals and semiconductors, which neglects this interaction, works as well as it does. It is difficult, however, to believe that the coloumb repulsion is the interaction responsible for superconductivity because there is no known way in which a repulsive interaction can give an energy gap. Furthermore, because the energy gap is very small, the interaction responsible for it must be very weak, much weaker than the coulomb interaction. The apparent lack of any mechanism for a weak attractive interaction was for some time the stumbling block in the way of any microscopic theory of superconductivity.

9.3.2. The electron–lattice interaction

Electrons may be represented by waves, and in an absolutely perfect crystal lattice free from thermal vibrations (i.e. cooled to absolute zero), these waves would propagate freely through the lattice without attenuation, rather in the way in which an electric wave can pass along a lossless periodic filter without attenuation. However, if the perfect periodicity of the lattice is destroyed by thermal vibrations, the lattice behaves like a periodic filter in which the values of some of the components (inductances or capacitors) fluctuate in a random manner. This causes a partial reflection of the wave, and in the same way an electron which encounters any departure from perfect periodicity of the crystal lattice would have a certain probability of being reflected, or scattered. We say that the electron *interacts* with the lattice and speak of the electron–lattice interaction. It is this electron–lattice interaction which determines the resistivity of pure metals and semiconductors at room temperature. Since both energy and momentum must be conserved when an electron is scattered, one of the vibrational modes of the lattice must be excited in the scattering process. This vibrational motion is quantized, so we speak of the emission (or absorption) of a phonon. A phonon bears the same relationship to a sound wave as a phonon does to a light wave.

An early step forward in the search for a microscopic theory came in 1950 when Fröhlich pointed out that the electron–phonon interaction was able to couple two electrons together in such a way that they behaved as if there was a direct interaction between them. In the interaction postulated by Fröhlich, one electron emits a phonon which is then immediately absorbed by another, and he was able to show that in certain circumstances this emission and subsequent absorption of a phonon could give rise to a weak attraction between the electrons of the sort which might produce an energy gap of the right order of magnitude. We may think of the interaction between the electrons as being *transmitted* by a phonon.†

We can represent the Fröhlich interaction schematically as in Fig. 9.2, where the straight lines represent electron paths and the wavy line represents a phonon. During the process of phonon emission, momentum is conserved, so we may write for the electron which emits the phonon

$$\mathbf{p}_1 = \mathbf{p}_1' + \mathbf{q}, \qquad (9.1)$$

where $\mathbf{p}_1$ is its momentum before scattering, $\mathbf{p}_1'$ its momentum after scattering, and $\mathbf{q}$ the momentum of the phonon which is given in magnitude by $q = h\nu_q/s$, where ν_q is the frequency of the phonon and s the velocity of sound. (This is analogous to the expression for the

---

† A macroscopic process in which there is an interaction between two particles resulting from the exchange of a third particle between them may be envisaged as follows. Suppose a skater throws a ball to a second skater. Then, due to conservation of momentum in the acts of throwing and catching, each skater will receive an impulse which tends to make him recede from his companion. There will be an apparent repulsion between the skaters despite the fact that there is no direct interaction between them. We could convert this into an attraction by replacing the ball by a boomerang, so that the first skater throws it in a direction away from his partner.

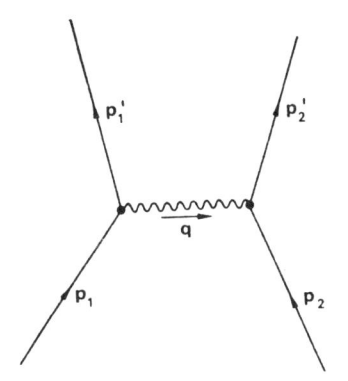

FIG. 9.2. Schematic representation of electron–electron interaction transmitted by a phonon.

momentum of a photon.) In the same way, when the phonon is absorbed by the second electron, the momentum of the latter changes from $\mathbf{p}_2$ to $\mathbf{p}_2'$ so that

$$\mathbf{p}_2 + \mathbf{q} = \mathbf{p}_2'. \tag{9.2}$$

From (9.1) and (9.2) we get

$$\mathbf{p}_1 + \mathbf{p}_2 = \mathbf{p}_1' + \mathbf{p}_2' \tag{9.3}$$

showing that momentum is conserved between the initial and final states, as we should expect. On the other hand, although energy is necessarily conserved between the initial and final states, it does *not* have to be conserved between the initial state and the intermediate state (i.e. the state in which the first electron has emitted a phonon but the second electron has not yet absorbed it), or between the intermediate state and the final state. This is because there is an uncertainty relationship between energy and time which takes the form $\Delta E \cdot \Delta t \simeq \hbar$. If the lifetime of the intermediate state Δt, is very short, there will be a large uncertainty, ΔE, in its energy, so that energy does not have to be conserved in the emission and absorption processes. Such processes, which do not conserve energy, are known as *virtual* processes, and virtual emission of a phonon is possible only if there is a second electron ready to absorb it almost immediately.

It turns out from the detailed quantum mechanics of the process that if $\varepsilon_1 - \varepsilon_1' < h\nu_q$, where ε_1 and ε_1' are the energies of the first electron before and after the virtual emission of the phonon, then the overall result of the emission and absorption processes is that there is an attraction between the two electrons. There is also, of course, the coulomb repulsion between the electrons; whether the net interaction is attractive or repulsive depends on whether the phonon-induced attraction exceeds the coulomb repulsion or vice versa.

Fröhlich's suggestion that the interaction responsible for superconductivity is one which involves lattice vibrations (or phonons) enabled him to predict the isotope effect before it had been discovered experimentally. The fact that an electron–phonon interaction is responsible for superconductivity also explains why superconductors are bad normal conductors. For example, lead, which has one of the highest critical temperatures, must have a fairly strong electron–phonon interaction and as a result is a poor conductor at room temperature, whereas the noble metals, gold and silver, which are very good conductors at room temperature, must be characterized by a weak electron–phonon interaction and do not become superconducting even at the lowest temperatures yet attained.

9.3.3. Cooper pairs

It is appropriate at this point to summarize the behaviour of conduction electrons in normal metals. For nearly all purposes it is permissible to neglect interactions between electrons in the normal state and to describe each electron as having individually an energy ε and momentum **p**. Because of the requirement that the wavefunction for an electron shall satisfy certain boundary conditions, there is a finite number of eigenstates in a given energy range, each eigenstate having an energy ε and momentum **p** as well defined as the uncertainty principle permits. The probability that a given eigenstate is occupied by an electron is then given by the Fermi–Dirac distribution

$$f(\varepsilon) = \frac{1}{e^{(\varepsilon-\varepsilon_F)/kT} + 1},$$

where ε_F is the Fermi energy. At absolute zero, the Fermi–Dirac function takes the form of a step-function, as shown in Fig. 9.3, and the points which represent the momenta of the electrons in three-dimensional momentum space occupy a sphere of radius p_F, known as the "Fermi sea", where

$$p_F = \sqrt{(2m\varepsilon_F)}.$$

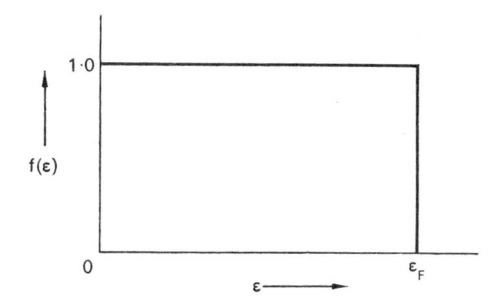

FIG. 9.3. Probability that a quantum state of kinetic energy ε is occupied by an electron, for the case of normal metal at absolute zero.

Following Fröhlich's discovery that the electron–electron interaction can be transmitted by phonons, the next step towards a microscopic theory of superconductivity was taken by Cooper,[†] who discussed what happens when two electrons are added to a metal at absolute zero so that they are forced by the Pauli principle to occupy states with $p > p_F$ as shown in Fig. 9.4. He was able to show that if there is an attraction between them, however weak, they are able to form a bound state so that their total energy is less than $2\varepsilon_F$. To see how this comes about, we will use some elementary ideas from quantum mechanics without attempting a rigorous derivation.

Consider first the case of two non-interacting electrons, with momenta $\mathbf{p}_1$ and $\mathbf{p}_2$. The two-electron wavefunction $\phi(x_1, y_1, z_1, \mathbf{p}_1, x_2, y_2, z_2, \mathbf{p}_2)$, which determines the probability that an electron with momentum $\mathbf{p}_1$ is at (x_1, y_1, z_1) while an electron with momentum $\mathbf{p}_2$ is at (x_2, y_2, z_2), is simply the product of two single-electron wavefunctions $\psi(x_1, y_1, z_1, \mathbf{p}_1)$ and $\psi(x_2, y_2, z_2, \mathbf{p}_2)$. We will take the dependence on the space coordinates as understood and write for brevity[‡]

$$\phi(\mathbf{p}_1, \mathbf{p}_2) = \psi(\mathbf{p}_1)\psi(\mathbf{p}_2).$$

The ψ's will be simply plane waves, or, more precisely, Bloch wavefunctions. Now if there is an interaction between pairs of electrons which causes scattering of the electrons accompanied by changes in their momenta, its effect is to "scramble" the wavefunctions so that the two-

[†] L. N. Cooper, *Phys. Rev.* **104**, 1189 (1956).

[‡] The reader who knows some quantum mechanics will realize that the right-hand side of this equation should properly be an antisymmetric sum of products of space- and spin-wave functions, but this is a complication which need not concern us on the level of this book.

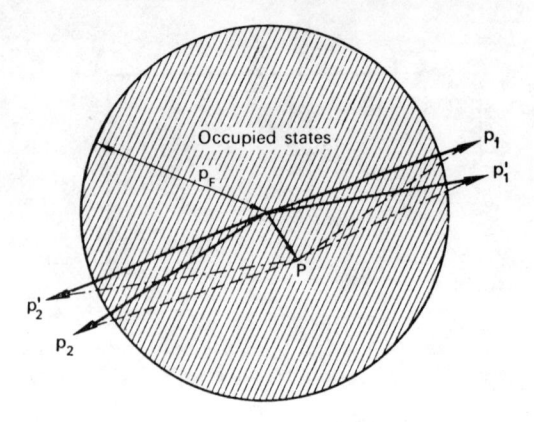

FIG. 9.4. Cooper's problem: two electrons interact above a filled Fermi "sea". The diagram is a section of momentum space showing the momenta of the electrons. The conduction electrons in a normal metal have momenta which are uniformly distributed within a sphere of radius p_F. Two additional electrons must have momenta $\mathbf{p}_1$ and $\mathbf{p}_2$ represented by points outside the sphere. Their resultant momentum $\mathbf{P}$ is conserved in scattering processes in which the individual momenta change from $\mathbf{p}_1, \mathbf{p}_2$ to $\mathbf{p}'_1, \mathbf{p}'_2$.

electron wavefunction is a mixture of wavefunctions comprising a wide range of momenta, and has the form

$$\Phi(x_1, y_1, z_1, x_2, y_2, z_2) = \sum_{i,j} a_{ij}\phi(\mathbf{p}_i, \mathbf{p}_j)$$
$$= \sum_{i,j} a_{ij}\psi(\mathbf{p}_i)\psi(\mathbf{p}_j). \qquad (9.4)$$

We can interpret the wavefunction Φ as meaning that the two electrons scatter each other repeatedly in such a way that their individual momenta are constantly changing, and that $|a_{ij}|^2$ gives the probability of finding the electrons at any instant with individual momenta $\mathbf{p}_i$ and $\mathbf{p}_j$. Since in each scattering event the total momentum of the two electrons is conserved, $\mathbf{p}_i$ and $\mathbf{p}_j$ must satisfy $\mathbf{p}_i + \mathbf{p}_j = \text{constant} = \mathbf{P}$. During the actual scattering process the electrons are subject to their mutual interaction, and if this is an attractive one, the potential energy which results from it is negative. Hence, over a period of time during which there are many scattering events, the energy of the two electrons is decreased by the time-average of this negative potential energy,† and the amount of this

† This is a pictorial description of a quantum-mechanical perturbation process. Strictly speaking, the perturbed wavefunction Φ is a stationary state, and we ought not to speak of time-averages. However, the description is good enough for our present purpose.

decrease is proportional to the number of scattering events which take place, i.e. to the number of ways in which we can choose two terms from the wavefunction Φ. It turns out to be a good enough approximation to assume that each scattering event contributes an equal amount $-V$ to the potential energy. (In quantum mechanical parlance, $-V$ is the "matrix element" of the interaction connecting two-electron states which have the same total momentum, and we are assuming that this is independent of the individual momenta of the electrons.)

We have said nothing in this section so far about the nature of the interaction, apart from requiring that it be attractive. If the interaction is of the sort described in § 9.3.2, arising from the virtual emission and absorption of a phonon, it turns out from the detailed theory that the probability of scattering is appreciable only if the energy deficit between the initial and intermediate states ($\varepsilon_1' + h\nu_q - \varepsilon_1$ in the notation of § 9.3.2) is small, i.e. if $\varepsilon_1 - \varepsilon_1' \simeq h\nu_q$. If we are considering a metal at absolute zero with two electrons added to it, all the eigenstates with kinetic energies up to ε_F are occupied, so both ε_1 and ε_1' must be above ε_F in order not to violate Pauli's principle.† The lowest values of ε_1 and ε_1' which are above ε_F and at the same time satisfy $\varepsilon_1 - \varepsilon_1' \simeq h\nu_q$ lie within an energy $h\nu_L$ of ε_F, where ν_L is an "average" phonon frequency typical of the lattice, about half the Debye frequency. Remembering that $\varepsilon = p^2/2m$, this limitation on the allowed values of ε_1 and ε_1' means that p_1 and p_1' must lie within a range $\Delta p = mh\nu_L/p_F$ of the Fermi momentum p_F. Since all the pairs of values of p_i and p_j which make up the wavefunction Φ must satisfy the condition $\mathbf{p}_i + \mathbf{p}_j = \mathbf{P}$, the allowed values of $\mathbf{p}$ can be found by the construction shown in Fig. 9.5. These momenta all begin or end in the ring whose cross-section is shaded. The number of such pairs is proportional to the volume of this ring and has a very sharp maximum when $\mathbf{P} = 0$, in which case the ring becomes a complete spherical shell of thickness Δp. Thus *the largest number of allowed scattering processes, yielding the maximum lowering of the energy, is obtained by pairing electrons with equal and opposite momenta.* It also turns out from the detailed quantum mechanical treatment that the matrix elements V are largest, and the lowering of the energy greatest, if the two electrons have opposite spins. This result comes from consideration of the spatial symmetry of the wavefunction and is analogous to the fact that the ground state of the hydrogen molecule also has opposite spins.

† Here and in what follows ε is used to denote the kinetic energy ($p^2/2m$). It does not include the potential energy which results from the phonon-induced interaction.

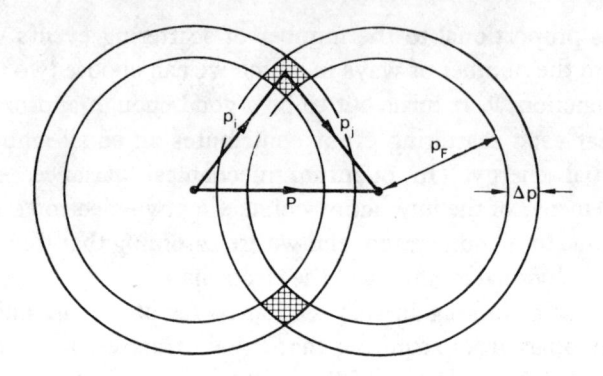

FIG. 9.5. The figure shows two shells of radius p_F and thickness $\Delta p = (mhv_L/p_F)$ whose centres are separated by the vector **P**. (The diagram has rotational symmetry about the vector **P**.) All pairs of momenta $\mathbf{p}_i$ and $\mathbf{p}_j$ which satisfy the relation $\mathbf{p}_i + \mathbf{p}_j = \mathbf{P}$ can be constructed as shown. The number of these pairs is proportional to the volume in **p**-space of the ring whose cross-section is shown shaded. This volume has a sharp maximum when $\mathbf{P} = 0$. (After Cooper.)

An essential condition for the wave function Φ to represent two electrons with the lowest possible potential energy is therefore that it be made up from wavefunctions of the form $\psi(\mathbf{p}\uparrow)\psi(-\mathbf{p}\downarrow)$, where the first term describes an electron with momentum $\mathbf{p}$ and spin up, and the second term an electron with momentum $-\mathbf{p}$ and spin down. Equation (9.4) now becomes

$$\Phi(x_1, y_1, z_1, x_2, y_2, z_2) = \sum a_i \phi(\mathbf{p}_i\uparrow, -\mathbf{p}_i\downarrow), \qquad (9.5)$$

where
$$\phi(\mathbf{p}_i\uparrow, -\mathbf{p}_i\downarrow) = \psi(\mathbf{p}_i\uparrow)\psi(-\mathbf{p}_i\downarrow)$$

and we have written a_i instead of a_{ii}. Such a wavefunction describes what is known as a *Cooper pair*. To obtain the total energy of the two electrons, we must add to their potential energy the total kinetic energy associated with their momenta. This is simply

$$W = 2\sum_i |a_i|^2 \, (p_i^2/2m).$$

Because the only states available to the two electrons are those with $p > p_F$, and also because $\Sigma |a_i|^2 = 1$ if Φ is normalized, the kinetic energy must exceed $2p_F^2/2m = 2\varepsilon_F$. The most important result from Cooper's analysis is that *in forming a pair with equal and opposite momenta, the lowering of the potential energy due to the interaction exceeds the amount by which the kinetic energy is in excess of $2\varepsilon_F$*. Hence, if the two electrons go into a state represented by the wavefunction Φ, in which they are continually scattered between states with equal and opposite momenta within the range $\Delta p = mhv_L/\varepsilon_F$, the total energy of the system is less than it would have been had they gone into states infinitesimally above p_F but with no interaction between them.

9.3.4. The superconducting ground state

The problem treated by Cooper is a somewhat unrealistic one in that it involves only two interacting electrons. In a metal there are about 10^{23} conduction electrons per cm^3, and while treating interactions between two electrons is clearly an improvement on a theory which ignores interactions altogether, one might reasonably ask why stop at two? Should we not take interactions between three or more electrons into account? The great step forward towards a microscopic theory of superconductivity came in 1957 when Bardeen, Cooper and Schrieffer† were able to show how Cooper's simple result could be extended to apply to many interacting electrons. The fundamental assumption of the Bardeen–Cooper–Schrieffer theory (usually known as the BCS theory) is that the only interactions which matter in the superconducting state are those between any two electrons which happen to make up a Cooper pair, and that the effect on any one pair of the presence of all the other electrons is simply to limit, through the Pauli principle, those states into which the interacting pair may be scattered, since some of the states are already occupied.

Cooper's result described in the previous section refers to what happens when two additional electrons are added to a metal at absolute zero. However, it applied equally well to the situation in which two electrons already belonging to the metal, with momenta infinitesimally below p_F, are transformed into a Cooper pair with equal and opposite momenta as described by the wavefunction Φ (9.5). The lowering of their potential energy due to their mutual interaction exceeds the amount by which their kinetic energy exceeds $2\varepsilon_F$. Hence, if we start with a metal at absolute zero, so that the distribution of the electrons is as in Fig. 9.3, we can form a state with lower energy by removing two electrons with p very slightly less than p_F and allowing them to form a Cooper pair. If we can do this for one pair we can do it for many pairs and lower the energy even further. This is possible because more than one pair of electrons can be represented by the same wavefunction Φ given by (9.5). In this case all the superconducting electrons can be

† J. Bardeen, L. N. Cooper, and J. R. Schrieffer, *Phys. Rev.* **108**, 1175 (1957).

represented together by a many-electron wave function Ψ_G which is a product of pair wavefunctions:†

$$\Psi_G(\mathbf{r}_1, \mathbf{r}_2, \ldots, \mathbf{r}_{n_s}) = \Phi(\mathbf{r}_1, \mathbf{r}_2)\Phi(\mathbf{r}_3, \mathbf{r}_4) \ldots \Phi(\mathbf{r}_{n_s-1}, \mathbf{r}_n), \quad (9.6)$$

where $n_s/2$ is the total number of pairs, $\mathbf{r}_n$ stands for the position coordinates (x_n, y_n, z_n) of the nth electron, and the Φ's on the right hand side *are the same for all pairs*. [It should be remembered that although, for clarity, we have not included the position coordinates on the right of (9.5), the ϕ's contain the position coordinates implicitly, as may be seen on reference to page 121. In fact, since the ψ's introduced on page 121 are plane waves proportional to exp $(i\mathbf{p} \cdot \mathbf{r}/\hbar)$, each ϕ on the right-hand side of (9.5) is a function of $(\mathbf{r}_{n-1} - \mathbf{r}_n)$.] The many-electron wavefunction (9.6) gives the probability of finding an electron at $\mathbf{r}_1$ while there is another at $\mathbf{r}_2$, and so on, irrespective of their momenta. The fact that we can write such a wavefunction, in which all the individual pairs are represented by (9.5), shows that there is no limit to the number of Cooper pairs which may be represented by wavefunctions of the form (9.5), and that we can regard the pair as a composite particle to which the Pauli principle in its simplest form does not apply; in other words, the pair may be regarded as a particle obeying Bose–Einstein statistics. This property of Cooper pairs, that they are all in the same quantum state with the same energy, will prove to be of great importance later on.

One might at first think that there is no limit to the number of electrons which may be raised from $p < p_F$ to form Cooper pairs with a resultant lowering of the total energy, so that we should end up with all the electrons having $p > p_F$! This, however, is clearly absurd, and the reason why is not hard to find. In order that a pair of electrons may be scattered from $(\mathbf{p}_i\uparrow, -\mathbf{p}_i\downarrow)$ to $(\mathbf{p}_j\uparrow, -\mathbf{p}_j\downarrow)$ the states $(\mathbf{p}_i\uparrow, -\mathbf{p}_i\downarrow)$ must first be occupied and the states $(\mathbf{p}_j\uparrow, -\mathbf{p}_j\downarrow)$ must be empty. As more and more electrons form Cooper pairs with $p > p_F$, the chance of finding the states $(\mathbf{p}_j\uparrow, -\mathbf{p}_j\downarrow)$ empty becomes progressively smaller and smaller, so the number of scattering processes which may take place is reduced, with a consequent decrease in the magnitude of the negative potential energy. Eventually a condition is reached in which the lowering of the potential energy is insufficient to outweigh the increase in the kinetic energy, and it is no longer possible to lower the total energy of the electrons by forming Cooper pairs. There will be an optimum arrangement which gives the lowest overall energy, and this arrangement can be described by

† The remarks on page 121 about antisymmetrization apply here also.

specifying the probability h_l of the pair state $(\mathbf{p}_l\uparrow, -\mathbf{p}_l\downarrow)$ being occupied in the wavefunction Ψ_G. This probability is related to, but is not the same as, the coefficient a_l which occurs in (9.5). The Pauli principle as applied to pairs requires that $h_l \leqslant 1$. According to the BCS theory, h_l is given by

$$h_l = \frac{1}{2}\left[1 - \frac{\varepsilon_l - \varepsilon_F}{\{(\varepsilon_l - \varepsilon_F)^2 + \Delta^2\}^{\frac{1}{2}}}\right], \tag{9.7}$$

where $\varepsilon_l = p_l^2/2m$ and the positive square root is taken. The quantity Δ, which has the dimensions of energy, turns out to be of fundamental importance and is given by

$$\Delta = 2h\nu_L \exp\left[-\{\mathcal{N}(\varepsilon_F)V\}^{-1}\right], \tag{9.7a}$$

where ν_L is the average phonon frequency introduced in § 9.3.3, $-V$ is the matrix element of the scattering interaction, and $\mathcal{N}(\varepsilon_F)$ is the density of states (ignoring spin) for electrons at the Fermi energy of the normal metal.

Figure 9.6 shows the probability h_l as given by the BCS theory, plotted as a function of p_l, when the wavefunction Ψ_G corresponds to the state of lowest overall energy, usually referred to as the ground state. Also shown in Fig. 9.6 is the probability of the single-electron state with momentum p_l being occupied in a normal metal at 0°K. The important feature of the figure is that *even at the absolute zero* the momentum distribution of the electrons in a superconductor does not show an abrupt discontinuity as in the case of a normal metal.

We thus see that the state of lowest energy (the ground state) occurs when all the electrons with momenta within a range $\Delta p = mh\nu_L/p_F$ about p_F are coupled together in Cooper pairs having opposite momentum and spin. This state is often referred to as a *condensed* state because the electrons are bound together to form a state of lower energy, as happens to the atoms of a gas when they condense to form a liquid. It is important to stress that the paired electrons described by the wavefunction Ψ_G must be regarded as all belonging to the same quantum state and having the same energy, because they are all continually being scattered between single-electron states having momenta within the range Δp, so that their modes of motion cannot be distinguished in any way whatsoever. The total energy of the interacting electron pairs is constant despite the fact that their momenta are continually changing, so the time-dependent factors by which each term on the right hand side of

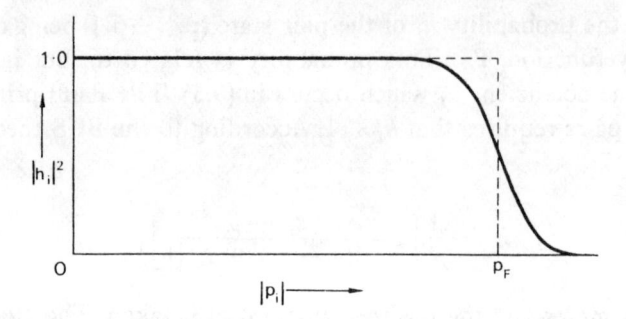

FIG. 9.6. ——— Probability h_i that the two-electron state $(\mathbf{p}_i\uparrow, -\mathbf{p}_i\downarrow)$ is occupied in the BCS ground-state wavefunction.
- - - - - Probability that a single-electron state with momentum $|\mathbf{p}_i|$ is occupied in a normal metal at absolute zero. Note that, in the superconducting ground state, even at absolute zero there are vacancies with $p_i < p_F$ and occupied states with $p_i > p_F$.

(9.5) must be multiplied to give a time-dependent wavefunction† all oscillate with the same frequency. Because these terms have the same frequency, they must have a definite phase relationship with each other, and so each α_i in (9.5) is in general complex. This is often expressed by saying that the superconducting ground state Ψ_G is a *coherent mixture* of single-electron wavefunctions $\psi(p_i)$.

9.3.5. Properties of the BCS ground state

Correlations

We saw in Chapter 6 that a superconducting-to-normal boundary is characterized by a length ξ of the order of 10^{-4} cm, called by Pippard the coherence length, which is the shortest distance within which there may be a significant change in the degree of order (or in the concentration of superelectrons). It can be shown that the single-pair wavefunction (9.5) has a spatial extent of about 10^{-4} cm, so that in a rather loose way we can regard the Cooper pair as a sort of large molecule having this size. It is natural, therefore, to identify the spatial extent of the pair wavefunction with the coherence length ξ.

A more precise definition of ξ can be given by posing the question: what is the probability of finding an electron with momentum $-\mathbf{p}$ and spin down in a volume element $d\tau_2$ at a distance r from a volume element

---

† All the wavefunctions mentioned hitherto have been time-independent. To obtain a time-dependent wavefunction each time-independent wavefunction must be multiplied by an oscillatory term $e^{-iEt/\hbar}$, where E is the total energy.

$d\tau_1$ which contains an electron with momentum **p** and spin up? In the normal metal there are no correlations, i.e. the probability is independent of r, and the answer to the question is $\frac{1}{4}n^2 d\tau_1 d\tau_2$, where n is the density of electrons. In the superconducting state the answer is again $\frac{1}{4}n^2 d\tau_1 d\tau_2$ for very large values of r, but for small values of r the probability is greater, showing that paired electrons are more likely to be close together than far apart (Fig. 9.7). This increased probability extends

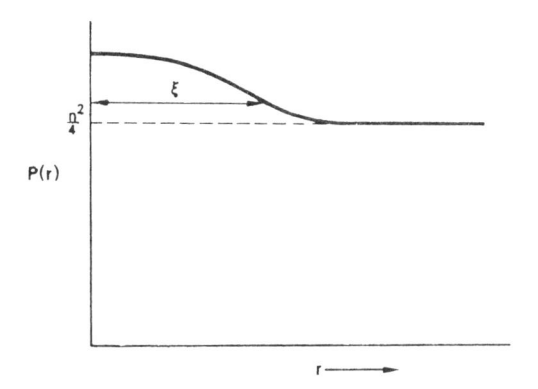

FIG. 9.7. The probability $P(r)d\tau_1 d\tau_2$ of finding an electron with momentum $-\mathbf{p}$ and spin down in a volume $d\tau_2$ at a distance r from a volume $d\tau_1$ containing an electron with momentum **p** and spin up. For the normal state (broken line) $P(r)$ is constant and equal to $n^2/4$, where n is the electron concentration. For the superconducting state $P(r)$ exceeds $n^2/4$ up to a range $\xi \sim 10^{-4}$ cm.

over a distance of about 10^{-4} cm in a pure type-I superconductor, and this distance is again identified with the coherence length. This confirms the interpretation of ξ as the spatial extent of the pair wavefunction given by (9.5). Notice that within a volume ξ^3 there lie the centres of mass of about 10^7 other pairs, so that the pair wave functions overlap considerably.

The energy gap

So far we have considered only the ground state of the superconductor, by which we mean the state of lowest energy, or the state of the superconductor at absolute zero. The next question we ask is: what happens if the superconductor is excited to a higher state, say by raising the temperature or by illuminating it with light of an appropriate wavelength? In § 9.1.1 we saw that the absorption of infrared radiation sets in quite abruptly for frequencies above a certain threshold frequency, and in § 9.2 we suggested that this might be because of the existence of an energy gap of some kind. We now show how this can be explained within the framework of the BCS theory.

If energy is imparted to a Cooper pair, say by letting light fall on it, we might think that this increase in energy would be brought about by increasing the magnitudes of the momenta which occur in the wavefunction Φ given by (9.5). However, Φ already contains a mixture of all values of momenta within a range $\Delta p = mhv_L/p_F$, subject only to the restriction that the total momentum is zero, so we cannot increase the energy of the pair simply by increasing the momenta of the electrons and at the same time maintaining the condition that their momenta are equal and opposite. What can happen, however, is that a pair may break up so that the electrons no longer have equal and opposite momenta. In this case they are unable to take part in such a large number of scattering events as when they formed a Cooper pair, and the negative potential energy resulting from their interaction is almost negligible. They behave almost like free electrons, and for this reason are referred to as "quasi-particles". It is not meaningful to talk of the momenta of the individual electrons *before* the pair was broken up because, as we have emphasized, the individual momenta of the electrons represented by the pair wavefunction Φ cannot be specified. It is, however, meaningful to talk of the momenta of the electrons *after* the pair is broken up, because they now behave almost as free electrons with well specified momenta. We may therefore ask how much energy is necessary to break up a pair so as to produce two electrons, or more properly quasi-particles, with momenta $\mathbf{p}_i$ and $\mathbf{p}_j$. (When we talk of quasi-particles of momenta $\mathbf{p}_i\uparrow$ and $\mathbf{p}_j\uparrow$ it must be understood that the complementary states $-\mathbf{p}_i\downarrow$ and $-\mathbf{p}_j\downarrow$ are empty, i.e. the quasiparticles have no partners with which to form Cooper pairs.) According to the BCS theory, the answer to the question is that the amount of energy required is

$$E = E_i + E_j = \{(\varepsilon_i - \varepsilon_F)^2 + \Delta^2\}^{\frac{1}{2}} + \{(\varepsilon_j - \varepsilon_F)^2 + \Delta^2\}^{\frac{1}{2}}, \quad (9.8)$$

where $\varepsilon_i = p_i^2/2m$ and the positive square roots are taken. The quantity Δ is given by (9.7a). Hence the minimum amount of energy required is 2Δ, which occurs when $p_i = p_j = p_F$ or $\varepsilon_i = \varepsilon_j = \varepsilon_F$. There is thus an energy gap of magnitude 2Δ in the excitation spectrum of the superconductor, and radiation of frequency ν is absorbed only if $h\nu > 2\Delta$.

This energy gap has two causes. Firstly, the splitting up of the pair so that the electrons no longer have equal and opposite momenta results in the disappearance of their binding energy, rather in the way that energy is necessary to split up a molecule into its constituent atoms. Secondly, if the state $\mathbf{p}\uparrow$ is occupied by an electron but the state $-\mathbf{p}\downarrow$ is empty, then the pair state $(\mathbf{p}\uparrow, -\mathbf{p}\downarrow)$ is not available to the remaining Cooper pairs, so that the number of scattering events in which they can participate is reduced, with a corresponding decrease in their binding energy. Hence the total energy of the entire system of electrons, which is what matters, is increased even further.

It is important to emphasize that by "a quasi-particle with momentum $\mathbf{p}_l$," is meant an electron in the state with momentum $\mathbf{p}_l\uparrow$ while the complementary state $-\mathbf{p}_l\downarrow$ is empty. The magnitude of $\mathbf{p}$ may be either greater than or less than p_F. In the ground state, where all electrons near p_F form Cooper pairs, the probability of a state with momentum p_l being occupied is given by h_l, as shown in Fig. 9.6. If, as a result of splitting up a pair, there is a quasi-particle in the state $\mathbf{p}_l$, then this state is now definitely occupied whereas previously the probability of its being occupied was h_l. If $p_l > p_F$, h_l is small, and we can say that after a pair is split up there is definitely an electron in the state $\mathbf{p}_l$, where previously it was most likely to be empty. We can therefore regard the quasi-particle in $\mathbf{p}_l$ as an electron. On the other hand, if $p_l < p_F$, h_l is close to unity, and the state $\mathbf{p}_l\uparrow$ is definitely occupied (and the state $-\mathbf{p}_l\downarrow$ definitely empty) where previously they were both most likely to be occupied. We can now regard the quasi-particle with momentum $\mathbf{p}_l$ as a vacancy or "hole" with momentum $-\mathbf{p}_l$. This view of a quasi-particle as an electron if $p_l > p_F$ or a hole if $p_l < p_F$ may seem rather artificial if $p_l \simeq p_F$, but if p_l is far removed from p_F the description is obvious.†

9.3.6. Macroscopic properties of superconductors according to the BCS theory

The critical temperature

The previous section was concerned with the ground state of the superconductor, except in so far as we considered what happened when a Cooper pair was split up into two quasi-particles, e.g. by radiation. If the temperature is raised above absolute zero, pairs are broken up by thermal agitation, and at any particular temperature the number of quasi-particles is given by the laws of statistical mechanics. There is one complication, however; the energy gap is not a constant but decreases as the

---

† The reader is warned that in the literature of superconductivity the term "hole" is used simply to denote an unoccupied state. It does not have the additional significance of a vacant state with negative effective mass near the top of an otherwise full band, as it does in semiconductor usage.

temperature rises. It is easy to see why this happens. As we saw in § 9.3.5, an electron in the state (p↑) without a partner in (−p↓) prevents the pair state (p↑, −p↓) from being available to Cooper pairs, and the pair interaction energy is diminished because the number of scattering events in which they may participate is lessened. This decrease in the pair interaction energy means a decrease in the energy gap. As the temperature rises, the number of quasi-particles increases and the energy gap continues to fall, until finally a temperature is reached at which the energy gap is zero. This is the critical temperature T_c, and above it the electrons cannot be represented by a correlated wavefunction of the type given by (9.5). The variation with T of the energy gap $E_g = 2\Delta$, as predicted by the BCS theory, is shown in Fig. 9.8; the shape of this curve has been

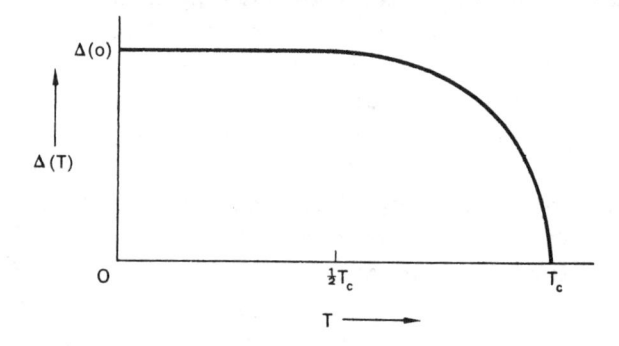

FIG. 9.8. Variation of Δ with temperature.

well confirmed by experiment. The theory also predicts that the critical temperature is simply related to the energy gap at absolute zero by

$$E_g(0) = 2\Delta(0) = 3 \cdot 5 k T_c, \tag{9.9}$$

where k is Boltzmann's constant. An experimental test of this relationship is given in Table 9.1, the values of $\Delta(0)$ being obtained from the infrared absorption measurements of Richards and Tinkham. The error is about ±0·20 in each case. The experimental values are close to the theoretical value, but the deviations are outside the experimental error; this can be put down to the simplifications made in the theory, such as the assumption that the matrix element V is independent of the change in momentum in the scattering process.

Inserting the expression (9.7a) for Δ into (9.9) gives an explicit equation for T_c:

$$3 \cdot 5 k T_c = 4 h \nu_L \exp\left[-\{\mathcal{N}(\varepsilon_F) V\}^{-1}\right]. \tag{9.10}$$

TABLE 9.1. RATIO OF SUPERCONDUCTING
ENERGY GAP AT ABSOLUTE ZERO TO kT_c

Superconductor	$2\Delta(0)/kT_c$ (Experimental)
Indium	4·1
Tin	3·6
Mercury	4·6
Vanadium	3·4
Lead	4·1

Since ν_L is proportional to $M^{-\frac{1}{2}}$, where M is the isotopic mass, this explains the origin of the isotope effect. The $M^{-\frac{1}{2}}$ dependence is frequently found in practice but is not universal; departures from it can be explained in terms of the effect of the Coulomb interaction between electrons.

The latent heat

Figure 9.8 shows that at temperatures below about $0\cdot6T_c$, the energy gap is substantially independent of temperature. A constant amount of energy, $2\Delta(0)$, is therefore needed to break up a Cooper pair and the number of pairs broken up at a temperature T in this range is proportional to $e^{-\Delta(0)/kT}$. This leads to an electronic specific heat which is proportional to $e^{-\Delta(0)/kT}$, as is found experimentally at low temperatures (cf. §§ 9.1.3 and 9.2 and Chapter 5).

At temperatures close to T_c the specific heat rises more rapidly with temperature because $\Delta(T)$ becomes smaller. Above T_c, where the electrons behave as in an ordinary metal, there is no contribution to the specific heat due to the splitting up of pairs, so at the critical temperature there is an abrupt fall in the specific heat as the temperature rises. The energy gap falls smoothly to zero as T approaches T_c, so the total energy of the electrons as T approaches T_c from below is exactly the same as when it approaches T_c from above. There is therefore no latent heat associated with the transition, and as a result no change in entropy. This combination of a discontinuity in the specific heat with the absence of a latent heat is characteristic of a second order transition; we can think of the critical temperature as the temperature at which the internal energy of the electrons *begins* to change (due to the appearance of the energy gap) rather than one at which it undergoes an abrupt change. This is in marked contrast to the case of a first-order transition such as the freezing of a liquid, where there is a discontinuity in the internal energy of the atoms accompanied by a latent heat.

The critical magnetic field

As we saw in Chapter 4, the critical magnetic field satisfies the relationship

$$\tfrac{1}{2}\mu_0 H_c^2 = g_n - g_s,$$

where g_n and g_s are the Gibbs free energy densities of the normal and superconducting phases. At absolute zero $g_n - g_s$ is simply equal to the difference between the internal energy densities in the normal and superconducting states (neglecting any difference in volume between the two phases), and according to the BCS theory this is simply the total binding energy of the Cooper pairs as compared with the normal metal in which no pairs are formed. This can be calculated and the result is

$$\tfrac{1}{2}\mu_0 H_0^2 = (g_n - g_s)_{T=0} = \tfrac{1}{2}\mathcal{N}(\varepsilon_F)[\Delta(0)]^2$$

But $\Delta(0)$ is related to T_c by $2\Delta(0) = 3 \cdot 5 k T_c$ so that

$$\frac{H_0^2}{T_c^2} = \frac{0 \cdot 47 \gamma}{\mu_0}, \tag{9.11}$$

where $\gamma = \tfrac{2}{3}\pi^2 \mathcal{N}(\varepsilon_F) k^2$ is the coefficient of T in the expression for the specific heat in the normal state (see Chapter 5 and § 9.1.3) and $\mathcal{N}(\varepsilon_F)$ is the density of states at the Fermi energy of the normal metal.

There is thus a law of corresponding states in the sense that if any two out of H_0, T_c, and γ are known, the third can be predicted. The accuracy with which relationship (9.11) is satisfied is limited by the accuracy of (9.9). It is important to note that there are no disposable parameters in (9.11), and it is one of the striking successes of the BCS theory that, despite the enormous variation in the properties of metals in the normal state, different superconductors do obey quite closely such a law of corresponding states.

The criterion for the existence of superconductivity

It is natural to ask whether *all* metals will exhibit superconductivity if cooled to low enough temperatures. The answer given by the BCS theory is that this is not necessarily so; metals will show superconducting behaviour only if the net interaction between electrons resulting from the combination of the phonon-induced and coulomb interactions is attractive. This is why the good normal conductors like silver and copper, which have a weak electron-phonon interaction, do not exhibit superconductivity (at any rate down to the lowest temperatures yet achieved).

9.3.7. The current-carrying states

We have said nothing so far about the phenomenon from which superconductivity acquired its name, viz., the disappearance of resistance. Both the ground state and the excited states which we have discussed have a perfectly isotropic distribution of electrons in momentum space, so that there are as many electrons travelling one way as the other, and no current flows. It is, however, possible to visualize a situation in which each Cooper pair, instead of having zero total momentum, has a resultant momentum **P** *which is the same for all pairs*. In this case the states which make up the pair wavefunction have momenta of the form

$$\left[\left(\mathbf{p}_i + \frac{\mathbf{P}}{2}\right)\uparrow, \left(-\mathbf{p}_i + \frac{\mathbf{P}}{2}\right)\downarrow\right]$$

instead of $(\mathbf{p}_i\uparrow, -\mathbf{p}_i\downarrow)$ as in (9.5). The entire momentum distribution is shifted bodily in momentum space by an amount $\mathbf{P}/2$, as shown in Fig. 9.9. The electron pairs are still able to take part in a large number of scattering process which conserve the total momentum. These can be described formally as scattering from $\left[\left(\mathbf{p}_i + \frac{\mathbf{P}}{2}\right)\uparrow, \left(-\mathbf{p}_i + \frac{\mathbf{P}}{2}\right)\downarrow\right]$ to $\left[\left(\mathbf{p}_j + \frac{\mathbf{P}}{2}\right)\uparrow, \left(-\mathbf{p}_j + \frac{\mathbf{P}}{2}\right)\downarrow\right]$. As seen by an observer moving with velocity $\mathbf{P}/2m$, the situation is indistinguishable from that already discussed in which the total momentum was zero, and the total energy of the electrons is the same except that it is increased by an additional kinetic energy $n_s P^2/8m$, where n_s is the total number of superconducting electrons. (We are considering here the situation at absolute zero, where there are no quasi-particles and all the electrons are paired.) The wavefunction of a Cooper pair now becomes†

---

† This separation of the pair wavefunction into a wavefunction Φ representing the relative motion of the electrons and a factor representing the motion of the centre of mass cannot always be done, but it seems to be valid under most conditions of interest.

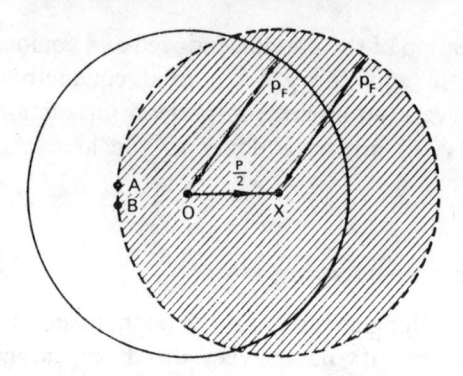

FIG. 9.9. Momentum distribution in current-carrying superconductor. The momentum vectors are uniformly distributed throughout a sphere of radius p_F (shown shaded) whose centre X is displaced by a vector $\mathbf{P}/2$ from the origin O. $\mathbf{P}$ is the total momentum of a Cooper pair. In the absence of a current the sphere is centred about the origin. Above a certain current density it is energetically possible for a pair to split up into two quasi-particles whose momenta are represented by the points A and B.

$$\Phi_P = \Phi e^{i\mathbf{P} \cdot (\mathbf{r}_1 + \mathbf{r}_2)/2\hbar}, \qquad (9.12)$$

where Φ is the pair wavefunction described by (9.5) and the exponential term represents the motion of the centre of mass of the pair with total momentum $\mathbf{P}$. If $\mathbf{r}$ denotes the position of the centre of mass of the pair, $\mathbf{r} = (\mathbf{r}_1 + \mathbf{r}_2)/2$, and (9.12) becomes

$$\Phi_P = \Phi e^{i\mathbf{P} \cdot \mathbf{r}/\hbar}. \qquad (9.12\mathrm{a})$$

In this picture the current is carried by pairs of electrons which have a total momentum $\mathbf{P}$. When a current is carried by an ordinary conductor, such as a normal metal or a semiconductor, resistance is inevitably present because the current carriers (either electrons or holes) can be scattered with a change in momentum so that their free acceleration in the direction of the electric field is hindered. This scattering may be due to impurity atoms, lattice defects or thermal vibrations. In the case of a superconductor, the electrons which make up a Cooper pair are constantly scattering each other, but since the *total* momentum remains constant in such a process there is no change in the current flowing. The only scattering process which can reduce the current flow is one in which the total momentum of a pair in the direction of the current changes, and this can only happen if the pair is broken up. However, this de-pairing requires a minimum amount of energy 2Δ, so the scattering can only happen if this energy can be supplied from somewhere. For low current densities there is no way in which this energy can be imparted to the pairs, so scattering events which change the total momentum of a pair are completely inhibited and *there is no resistance.*†

The momentum of the pairs is related to the current density **j** by $\mathbf{j} = en_s\mathbf{P}/2m$, where n_s is the total number of superconducting electrons and e the electronic charge. As **j** increases, the momentum distribution shown in Fig. 9.9 becomes more and more displaced, until finally it becomes energetically possible for a Cooper pair to split up into two quasi-particles whose momenta are represented by points such as A and B on the surface of the displaced sphere nearest to the origin. This can be seen if we take as the zero of energy the energy of the superconductor with a full complement of pairs (no quasi-particles) and no current (**P** = 0). Relative to this zero, the energy of the superconductor carrying a current, but still with no quasi-particles, is

$$W_1 = n_s P^2/8m$$

which is the additional kinetic energy of n_s electrons each with momentum **P**/2.

If the superconductor carries no current but has one pair split up, the minimum excitation energy required is 2Δ, and this occurs when the quasi-particles have momenta p_F [see (9.8)]. Suppose these quasi-particles have momentum vectors pointed to the left (Fig. 9.9) and that the whole distribution is now displaced an amount $P/2$ to the right so that the momenta of the quasi-particles are represented by A and B. The energy of the superconductor relative to our zero is now

$$W_2 = 2\Delta + (n_s - 2)\frac{P^2}{8m} + 2\left\{\frac{(p_F - P/2)^2}{2m} - \frac{p_F^2}{2m}\right\},$$

where the first term is the energy required to break up the pair, the second term the additional kinetic energy of the $(n_s - 2)$ paired electrons, and the third term the change in kinetic energy of the quasi-particles. It will be energetically favourable for the pair to split up if $W_1 > W_2$, i.e. if

$$\frac{p_F P}{m} > 2\Delta \quad \text{or} \quad P > \frac{2m\Delta}{p_F}. \tag{9.13}$$

---

† It is clear that this explanation of zero resistance cannot apply to the "gapless" superconductors referred to on page 116. It is believed that in this special category of superconductors scattering of pairs is inhibited not by the presence of an energy gap but by the strongly correlated nature of the pair wavefunction. If this is correct, it seems certain that such a mechanism must also be important for those superconductors which do exhibit an energy gap.

Since P is proportional to the current density this means that there will be a critical current density above which scattering accompanied by a change in total momentum (i.e. resulting in the breaking up of a pair) may take place. Above this critical current density resistance will appear. Combining the expression for j with the condition (9.13) we find

$$j_c = \frac{en_s\Delta}{p_F} \quad (9.14)$$

In Chapter 8 we stated that for tin, the critical current density at absolute zero is about 2×10^7 A cm^{-2}. In (9.14) the most uncertain quantity is n_s, and if we substitute for Δ the value $1 \cdot 80kT_c$ (Table 9.1) and for p_F the value corresponding to $v_F = 6 \cdot 9 \times 10^7$ cm s^{-1}, we find $n_s = 8 \times 10^{21}$ cm^{-3}. This is appreciably less than one electron per atom, but is not unreasonable in view of the complicated band structure of tin.

At a non-zero temperature, some of the pairs will be broken up into quasi-particles even for currents below the critical current. The quasi-particles behave very much like normal electrons; they can be scattered or excited further, and if they carry a current, will exhibit resistance. On the other hand, the remaining pairs retain the properties of the electrons at absolute zero, and cannot be scattered unless an amount of energy 2Δ is available. These paired electrons are what we have referred to as "superelectrons". Thus we can identify two almost independent fluids of normal and superelectrons, a concept made use of in Chapter 1.

9.3.8. The pair wavefunction: long-range coherence

We saw in the previous section that if a Cooper pair has a total momentum it can be presented by a wavefunction

$$\Phi_P = \Phi e^{i\mathbf{P}\cdot\mathbf{r}/\hbar} \quad (9.12a)$$

where Φ is the wavefunction given by (9.5). We also saw in § 9.3.4 that Φ is characterized by a coherence length $\xi(\sim 10^{-4}$ cm$)$, which is the distance within which there is a spatial correlation between electrons with equal and opposite momenta. If we regard the pair as being in a rather loose sense a sort of bound molecule, then Φ represents the internal motion of this molecule and ξ its spatial extent. The exponential term in (9.12a) is a travelling wave which represents the motion of the centre of mass of the pair, and the wavelength of this wave is h/P, which is the de Broglie wavelength of a particle of momentum P. The phase coherence of this travelling wave extends over indefinitely large distances, very much greater than ξ, in fact the wave travels unscattered through the whole volume of the superconducting metal. In the case of a persistent current in a ring, therefore, the phase coherence extends over centimetres; in the case of the winding of a superconducting solenoid over miles! The consequences of this very long range coherence are discussed in Chapter 11.

The energy gap in the electron energy spectrum of a superconductor can be reduced by a number of agencies; for example, incorporation of magnetic impurities. It is found that magnetic impurities lower the transition temperature as well as reducing the energy gap, but that the transition temperature can still be above zero at an impurity concentration which has reduced the energy gap to zero. In this condition the metal is resistanceless, implying long-range coherence of the electron-pair wave, even though there is no energy gap. This condition is called *gapless superconductivity*; it implies that it is the long-range coherence of the electron-pair wave, not the energy gap in the electron energy spectrum, which is the essential feature of superconductivity. In gapless superconductors there is at the Fermi level a minimum in the density of states, but no actual gap. In general, if a perturbation causes a superconductor to pass into the normal state through a second-order transition, the superconductor goes into the gapless state before it becomes normal. A discussion of gapless superconductivity is given in the articles by Meservy and Schwartz and by Maki.†

---

† These articles are in *Superconductivity*, ed. R. D. Parks, 1969 (Marcel Dekker Inc., New York).

CHAPTER 10

TUNNELLING AND THE ENERGY GAP

WE DESCRIBE in this chapter a technique introduced by Giaever in 1960 for the direct measurement of the energy gap in a superconducting metal by studying the tunnelling of electrons into the metal through a very thin insulating film. The discovery of this technique soon after the publication of the BCS theory made a great contribution towards our understanding of superconductivity because of the ease with which it enabled energy gaps to be measured experimentally.

10.1. The Tunnelling Process

To understand the method, consider first what happens if two plates made from a normal metal are separated by a very thin gap. The energy band diagram for this situation is shown in Fig. 10.1. Under ordinary

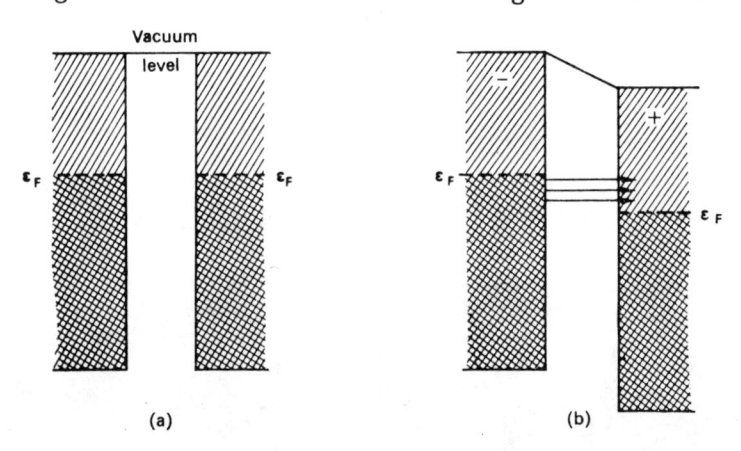

FIG. 10.1. Tunnelling between normal metals. The single-hatching represents empty states and the double hatching occupied states. At absolute zero, with no bias between the plates, as in (a), tunnelling is completely forbidden by the Pauli principle. If a positive voltage is applied to the right-hand plate, as in (b), so that the Fermi levels no longer coincide, there are occupied states on the left opposite to unoccupied states on the right, and tunnelling can take place as shown by the arrows.

conditions an electron in either plate cannot leave the metal because its energy is considerably less than the potential energy of a free electron in the vacuum outside. But if the separation between the metals is extremely small, an electron in one metal can cross to the other by virtue of a quantum-mechanical phenomenon known as tunnelling. In this process the electron is represented outside the metal by an exponentially attenuated standing wave whose amplitude falls off as $e^{-x/\zeta}$ instead of the usual travelling wave which represents the electron inside the metal. The length ζ is typically of the order of 10^{-8} cm, so the wave is attenuated very quickly indeed. However, if the gap is very thin (of the order of 10^{-7} cm) there is a small but significant chance that an electron may pass through the potential barrier separating the plates, and exchange of electrons between the two metals becomes possible. This "tunnelling" process is also possible if the two electrodes are separated by a very thin insulating film, such as the film of oxide which often forms on the surface of metals exposed to the atmosphere, and this is the situation which is important in practice.

There are two conditions which must be fulfilled for tunnelling to take place, apart from the obvious one that the separation between the metals must not be large compared with the attenuation length ζ of the tunnelling wavefunction. First, energy must be conserved in the process, i.e. the total energy of the system, including the metals on both sides of the insulating film, must be the same before and after tunnelling. Second, tunnelling can take place only if the states to which the electrons tunnel are empty, otherwise the process is forbidden by the Pauli principle. To take as an example the situation shown in Fig. 10.1a, in which we have for simplicity assumed that the metals are separated by a vacuum, at absolute zero there could be no tunnelling from one metal to the other because all the states which satisfy the first condition are occupied on both sides. However, if there is a small voltage difference between the metals (say the left-hand one negative with respect to the right), the energy levels will be shifted with respect to each other as in Fig. 10.1b. There will now be empty states on the right opposite to the topmost occupied states on the left, and tunnelling can take place from left to right. The number of states uncovered in this way is proportional to the voltage difference, and if the tunnelling probability is constant, as it is for very small bias voltages, the resulting current is linearly proportional to the voltage.

With superconductors, a number of situations may be envisaged. We may have one of the electrodes superconducting and the other normal, or both electrodes made of the same superconductor, or the electrodes made from two different superconductors. All of these have current–voltage characteristics which are peculiar to the particular combination of metals involved.

10.2. The Energy Level Diagram for a Superconductor

The difference between tunnelling involving superconductors and tunnelling involving normal metals can be explained in terms of the difference between their energy level diagrams. When we draw a band diagram for an ordinary metal, as in Fig. 10.1, we are showing the range of energies allowed to an individual electron. The electrons are independent of each other, so the energy of a particular electron is not affected by whether or not another level happens to be occupied. In the case of a superconductor this is no longer true. The electrons in the condensed

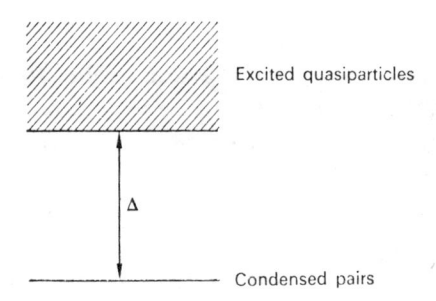

FIG. 10.2. Energy level diagram for superconductor.

state are not independent of each other, and their contribution to the total energy depends very much on whether or not they have a partner with equal and opposite momentum. We cannot therefore draw a band diagram in the usual way. As we have already pointed out, *all* the pairs have the same energy because they are all represented by the same wavefunction Φ [eqn. (9.5)], and we can therefore draw a single level as in Fig. 10.2 to represent the average energy *per electron* (or one half the energy of a Cooper pair) in the condensed state. This level can only contain paired electrons. As we saw in § 9.3.2, this level can contain many pairs because the Pauli principle in its usual form does not apply to Cooper pairs. If any of the pairs is split up, the energy of the system is increased in accordance with (9.8), namely

$$E_i + E_j = \{(\varepsilon_i - \varepsilon_F)^2 + \Delta^2\}^{\frac{1}{2}} + \{(\varepsilon_j - \varepsilon_F)^2 + \Delta^2\}^{\frac{1}{2}},$$

so that each of the resulting quasi-particles can be regarded as contributing an amount of energy $\{(\varepsilon - \varepsilon_F)^2 + \Delta^2\}^{\frac{1}{2}}$, where $\varepsilon = p^2/2m$. Since the quasi-particles behave almost like independent electrons, their allowed energy values can be represented by a continuum of levels separated by an interval Δ from the level representing the pairs, as in Fig. 10.2. (Notice that the quantity we have called the energy gap is 2Δ. This is because if quasi-particles are produced by splitting up a pair they are always produced two at a time. It is, however, possible to inject a single quasi-particle by tunnelling, and if this is done the minimum energy added to the system is Δ.)

At absolute zero there are no quasi-particles and the continuum states are empty. At temperatures above absolute zero, the continuum levels are partly filled in accordance with the laws of statistical mechanics.

10.3. Tunnelling Between a Normal Metal and a Superconductor

Consider first tunnelling between a normal metal and a superconductor at absolute zero. The energy level diagrams are as shown in Fig. 10.3. With the plates at the same potential, the Fermi level in the normal metal coincides with the level representing the condensed pairs in the superconductor as in Fig. 10.3a. Again tunnelling is impossible if there is no potential difference between the plates.

If a positive bias of V volts is applied to the superconductor, all its energy levels are lowered† relative to those of the normal metal, but no tunnelling processes which conserve energy are possible until V reaches the value Δ/e, when the bottom of the continuum of quasi-particle levels coincides with the Fermi level in the normal metal, as shown in Fig. 10.3b. It now becomes possible for electrons in the normal metal to tunnel into the quasi-particle states, and the number of electrons which may tunnel increases steadily as the bias increases, so that the resulting current increases monotonically with V as shown in Fig. 10.3d. If the superconductor is given a negative bias voltage, so that all its energy levels are raised relative to those of the normal metal, no tunnelling can occur until the voltage becomes equal to $-\Delta/e$, when a completely new process becomes possible which involves the splitting up of a Cooper pair. It should be appreciated that if only one electron is involved in a

---

† An increase in electrostatic potential lowers the potential energy because the charge on the electrons is negative.

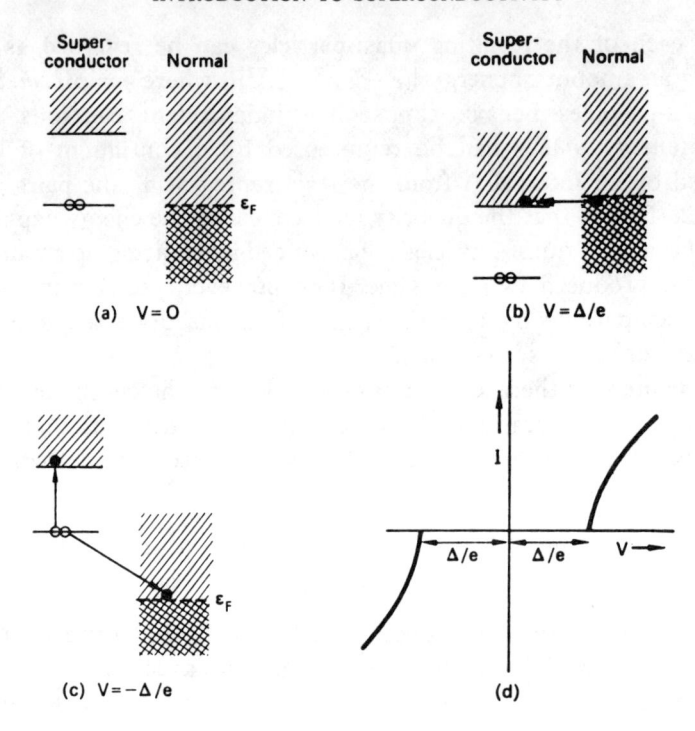

FIG. 10.3. Tunnelling between a normal metal and a superconductor. (a), (b), and (c) show the energy levels at 0°K for various differences in voltage V between a superconductor and a normal metal. V is taken as positive if the superconductor is biased positively relative to the normal metal. The singly-hatched areas denote empty states and the double-hatched areas denote occupied states. (d) Shows the resulting current–voltage characteristic.

transition the only processes which can conserve energy are those in which the electron moves horizontally on the energy band diagram between states of the same energy. However, if two electrons are involved it becomes possible for one to gain energy and the other to lose it as long as the total energy is conserved. Such a process is shown in Fig. 10.3c. In this, a Cooper pair splits up, one of the electrons tunnelling to an empty state just above the Fermi level of the normal metal with a loss of energy Δ, while the second electron, having lost its partner, is converted into a quasi-particle and occupies the lowest excited state in the superconductor with a gain in energy Δ. Thus the total energy of the complete system before the transitions indicated by arrows have taken place is equal to the total energy after, and the process is allowed. The number of pairs that may split up in this way increases with the bias

TUNNELLING AND THE ENERGY GAP 145

voltage, because more quasi-particle states and states in the normal metal become accessible, and the resulting negative current increases with $-V$ as shown in Fig. 10.3d. Δ is thus given directly by the voltage at which the tunnelling current between the superconductor and any normal metal shows a sudden increase. At temperatures above absolute zero, a very small current may flow at voltages between the values $\pm\Delta/e$, because there are a few electrons excited to states above the Fermi level in the normal metal which may, if there is positive bias, tunnel into the quasi-particle states, and there are a few empty states below the Fermi level of the normal metal to which one of the members of a pair may tunnel if there is negative bias. However, there will still be a sharp rise in tunnelling current when $V = \pm\Delta/e$. Therefore, from the tunnelling $I - V$ characteristic we can directly determine the energy gap ($= 2\Delta$) of the superconductor. Tunnelling is a relatively easy experiment and is a most useful way of measuring the energy gap.

10.4. Tunnelling Between Two Identical Superconductors

The energy level diagram for two identical superconductors with no applied bias is as shown in Fig. 10.4, where it is assumed that the temperature is above absolute zero so that the quasi-particle states are partially occupied. It is possible for quasi-particles to tunnel in either direction, because the states to which they may tunnel are not completely full, but at zero bias the current due to tunnelling from left to right will be the same as that from right to left, so that no net current flows.

Now suppose the left-hand electrode is biased positively by a voltage V relative to the right-hand electrode. This means that the energy level diagram on the left will be shifted downwards relative to that on the right by an amount eV, as shown in Fig. 10.4b. There will now be a net flow of electrons from right to left, because the lowest quasi-particles on the left have no states on the right to tunnel into, while all the quasi-particles on the right are able to tunnel. The current increases with V until the occupied quasi-particle states on the left are below the bottom of the continuum states on the right, so that quasi-particles can no longer tunnel from left to right. Since the quasi-particles will all lie within about kT of the lowest level, this stage is reached when $eV \simeq kT$ or $V \simeq 10^{-4}$ V. If V is increased further the current remains more or less constant because none of the quasi-particles on the left can tunnel, and those on the right, which are able to tunnel, are constant in number. However, when V reaches the magnitude $2\Delta/e$, an additional process

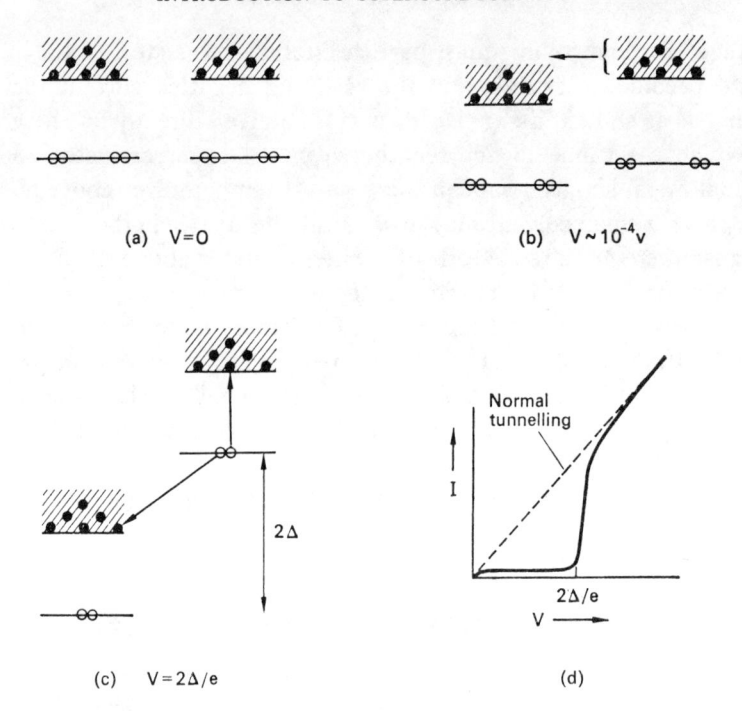

FIG. 10.4. Tunnelling between identical superconductors. (a), (b), and (c) energy levels. ⊖⊖ denotes a Cooper pair, ● denotes a quasi-particle. (d) Current–voltage characteristic.

becomes possible which involves the splitting up of a Cooper pair, as shown in Fig. 10.4c. One of the electrons tunnels into the left-hand superconductor so as to occupy the lowest quasi-particle state. This process is accompanied by a loss of energy Δ. The second electron, having lost its partner, is converted into a quasi-particle and occupies the lowest excited state on the right with a gain in energy Δ. Thus the total energy before the transitions indicated by the arrows have taken place is identical with the total energy after, and the process is allowed. As a result of this process, extra electrons flow from right to left and the current increases. If V is increased beyond $2\Delta/e$, a process of this sort continues to be possible except that the quasi-particles do not go into the lowest excited states on each side. There is now a large number of combinations of excited states which can serve as final states, and the tunnelling current increases rapidly, as shown in Fig. 10.4d. In this case also we can determine the energy gap from the voltage at which the tunnelling current increases rapidly.

10.5. The Semiconductor Representation

The reader is warned that there is another way of representing energy levels in a superconductor which is widely used in the literature. In this model the energy levels are represented by two bands, one of which is completely full at absolute zero and the other completely empty, as in a semiconductor. The width of the gap between the bands is 2Δ, not Δ as in Fig. 10.2, though, as we shall see, the two representations are equivalent.

To see how this model arises, consider a superconductor at a temperature above absolute zero, so that there are quasi-particles present, and suppose a single electron is added to it. Then one of two things may happen. The added electron may go into a momentum state $\mathbf{p}\uparrow$ whose partner $-\mathbf{p}\downarrow$ is empty. In this case it behaves as an excited quasi-particle. Alternatively, it may join up with another unpaired electron and form a Cooper pair. Compared with the first case, the second gives rise to a total energy which is lower by at least 2Δ. Since the superconductor is in the same condition before the electron is added in both instances, the energy of the extra electron before it enters the superconductor must be lower by at least 2Δ in the second case if energy is to be conserved overall. We may therefore regard the added electron as belonging to one of two bands which are separated in energy by 2Δ. An electron can only enter the superconductor with overall conservation of energy if, *before entering*, its energy lies above the bottom of the upper band or below the top of the lower band. These bands therefore define the ranges of energy which an electron must have if it is to tunnel into the superconductor. It is important to realize that in this model both the bands are *single electron* bands. It should not be thought that the levels in the upper band correspond to quasi-particles with $p > p_F$ and the lower to quasi-particles with $p < p_F$; a full range of momentum values is present in each case.

The semiconductor representation is most easily illustrated by considering tunnelling between a normal metal and a superconductor. In Fig. 10.5a,b we describe this process in terms of the excited quasi-particle representation, as we did in § 10.3 and Fig. 10.3. In Fig. 10.5c,d a description is given in terms of the semiconductor representation. The physics of the process is much more clearly brought out by the quasi-particle representation. The advantage of the semiconductor representation is that the allowed transitions always correspond to horizontal arrows which represent transitions made by a single electron, and this

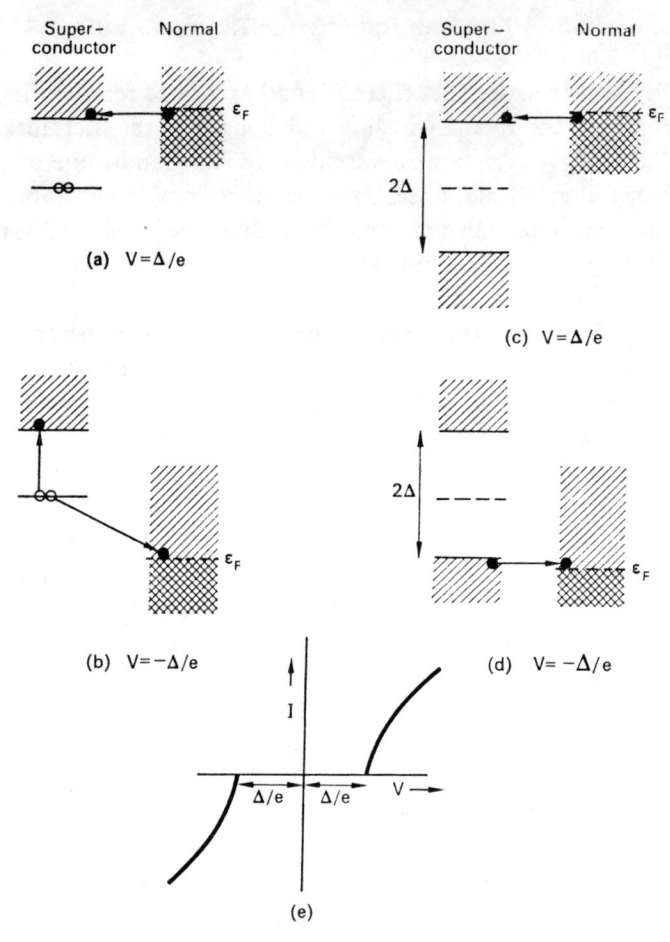

FIG. 10.5. Tunnelling between a superconductor and a normal metal at 0°K. (a) and (b) Excited quasi-particle representation. (c) and (d) Semiconductor representation. (e) Current–voltage characteristic.

makes the detailed interpretation of tunnelling phenomena somewhat easier; probably for this reason, it has been much more commonly used in the literature of tunnelling. It will be noted that an electron can only be injected into the lower band if there is a quasi-particle with which it can combine to form a Cooper pair, so the number of empty states in the lower band is equal to the number of quasi-particles in the upper band, rather like an intrinsic semiconductor. The mechanism by which the two bands come about is, however, totally different from that responsible for the band structure of a semiconductor and great care must be used in applying this representation in the case of a superconductor. The difference between the two representations has been discussed by Schrieffer[†] and Adkins.[‡]

10.6. Other Types of Tunnelling

There are other, more complicated, types of tunnelling. For example, if the two metals are dissimilar superconductors, a characteristic as shown in Fig. 10.6 is obtained. There is a negative resistance region between $V = (\Delta_2 - \Delta_1)/e$ and $V = (\Delta_2 + \Delta_1)/e$. The explanation of this negative resistance region depends on the way in which the density of states in the quasi-particle band varies with energy, and we shall not discuss it here. For further details the reader is referred to the original paper by Giaever and Megerle.[§]

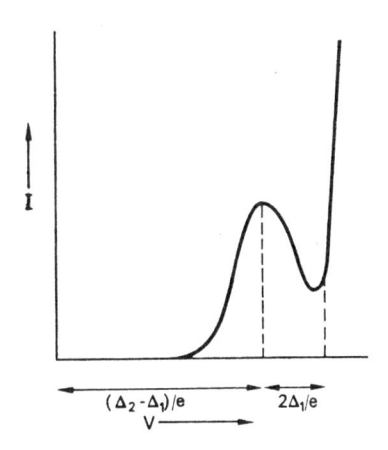

FIG. 10.6. Tunnelling between two superconductors with energy gaps Δ_1 and Δ_2 ($\Delta_2 > \Delta_1$).

There is also the possibility, in the case of two superconductors, of two electrons which form a pair tunnelling *as a pair*, so that they maintain their momentum pairing after crossing the gap. This type of tunnelling, known as Josephson tunnelling, is possible because the superconducting ground state can contain many pairs. It only appears under very special circumstances, namely exceptionally thin insulating layers ($<10^{-7}$ cm). Josephson tunnelling, the consequences of which are discussed at length in the following chapter, takes place when there is no

[†] J. R. Schreiffer, *Rev. Mod. Phys.* **36**, 200 (1964).
[‡] C. J. Adkins, *Rev. Mod. Phys.* **36**, 211 (1964).
[§] I. Giaever and K. Megerle, *Phys. Rev.*, **122**, 1101 (1961).

difference in voltage between the superconductors, so that a current may flow without any accompanying voltage drop. We might call it a tunnelling supercurrent. This supercurrent has a critical current density j_c which is characteristic of the junction.

If the Josephson tunnelling current density exceeds the value j_c, a voltage difference V appears across the junction and two processes occur. Some electrons tunnel individually, as in § 10.4, with an I–V characteristic as shown in Fig. 10.4d or Fig. 10.6 (depending on whether the superconductors are identical or not) and, at the same time, some electrons continue to tunnel in the form of electron-pairs. However, the condensed states are no longer opposite each other on an energy-level diagram, so pairs cannot tunnel from one condensed state to the other with conservation of energy if the energy of the electron pairs alone is considered. The energy balance is made up by the emission of a photon of electromagnetic radiation of frequency v such that

$$hv = 2eV. \qquad (10.1)$$

The factor 2 can be considered as arising either because the pair can be considered as a particle with charge $2e$, or because two electrons each with charge e are involved in the transition. Whichever way one looks at it, the energy required to make up the balance is $2eV$. This process, which involves the emission of radiation, is known as the a.c. Josephson effect. Since V is normally of the order of 10^{-3} V, the radiation is in the short wavelength microwave part of the spectrum. The emission of such radiation from very thin tunnel junctions has been detected by Langenburg, Scalapino, and Taylor.

Josephson tunnelling will be considered more fully in the next chapter (§ 11.3.1 *et seq.*).

10.7. Practical Details

Tunnelling is a very useful phenomenon, because we can use it to measure simply and directly the energy gap of a superconductor. The majority of superconducting tunnelling experiments have been carried out using evaporated films of the two metals. In a typical experiment a thin film of one of the metals is evaporated on to a glass plate (such as a microscope slide) in the shape of a strip a millimetre or so wide. This film is then oxidized by exposure to air or oxygen until a layer of oxide a few tens of Ångstrom units thick is built up on the surface. A film of the second metal is then evaporated, usually as a strip which crosses the first one at right angles, so that the area through which tunnelling can take place is a few square millimetres. Electrical contact is then made to the films, as a rule with indium solder, so that the current–voltage characteristic can be observed. Once the I–V characteristic has been observed, the energy gap can be determined from the voltage at which the curve shows a pronounced change in slope. This occurs at a voltage equal to Δ/e for superconductor–normal tunnelling and to $2\Delta/e$ for superconductor–superconductor tunnelling (see Figs. 10.3d and 10.4d).

Tunnelling currents are usually less than 10^{-3} A for applied voltages around 10^{-3} V, so care must be taken with the electrical measurements. It is common practice to superpose a small alternating component on the steady voltage V applied to the sandwich; the current then contains an alternating component which is proportional to the value of the differential conductance dI/dV. This a.c. component can be amplified, using a tuned amplifier. In this way a very sensitive direct measurement can be made of the slope of the I–V curve.

Figure 10.7 shows some measurements of the temperature dependence of the energy gap of indium, tin and lead, measured by Giaever and

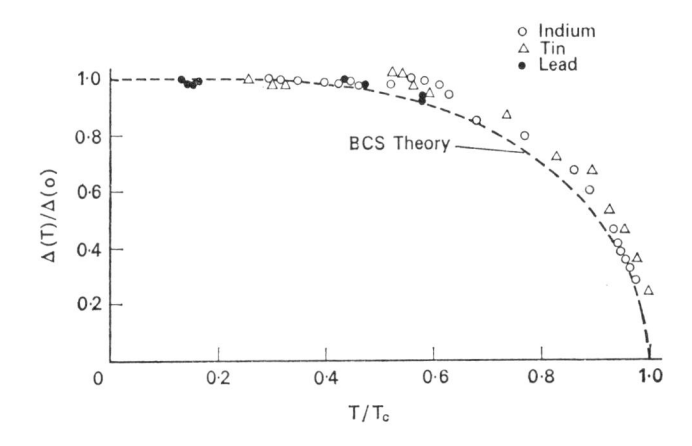

FIG. 10.7. The energy gap of lead, tin, and indium versus temperature, as determined by tunnelling experiments. (After Giaever and Megerle.)

Megerle who studied tunnelling from these metals into a normal metal. The fit with the temperature dependence predicted by the BCS theory is shown to be quite good. We should not expect the BCS theory to make *absolute* predictions of the energy gap and transition temperature of a superconductor with any great accuracy, so the figure shows the ratio of the energy gap at temperature T to the energy gap at 0°K, plotted against T expressed as a fraction of the transition temperature T_c. The curve obtained in this way should be the same for all superconductors.

One of the most impressive examples of the power of the tunnelling technique in measuring energy gaps is its use to measure the dependence of the energy gap on an applied magnetic field. As we saw in Chapter 5, the superconducting to normal transition in the presence of a magnetic field is a first-order transition, which implies that the energy gap should go abruptly to zero as the specimen is driven normal by the field. The data represented by circles in Fig. 10.8, which show the energy gap of an

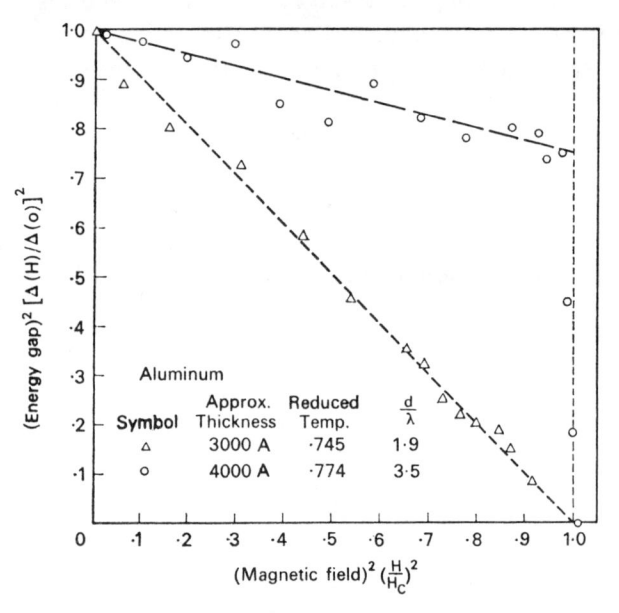

FIG. 10.8. Energy gap of aluminium as a function of magnetic field for films of thickness 3000 Å and 4000 Å. (After Douglass.)

aluminium film 4000 Å thick, obtained from tunnelling into superconducting lead, show the expected behaviour. The small decrease of energy gap with increasing field before H_c is reached can be explained in the same way as the field-dependence of the penetration depth discussed in §8.4. However, for a film which is 3000 Å thick, represented by triangles in Fig. 10.8, the energy gap falls linearly to zero as H increases to H_c. This gradual disappearance of the energy gap is indicative of a second-order transition, and is effective confirmation of the prediction by Ginzburg and Landau, mentioned in § 8.5, that below a certain film thickness the transition in a magnetic field should become of second order.

CHAPTER 11

COHERENCE OF THE ELECTRON-PAIR WAVE; QUANTUM INTERFERENCE

11.1. Electron-pair Waves

WE HAVE seen that in a superconductor the resistanceless current involves the motion of pairs of electrons. When considering the current, each pair may be treated as a single "particle" of mass $2m$ and charge $2e$ whose velocity is that of the centre of mass of the pair. As with ordinary particles, these current carriers may be described by means of a wave. In a normal metal the conduction electrons suffer frequent scattering accompanied by violent changes of phase and so their electron waves are coherent only over very short distances. The Cooper pairs in a superconductor are not, however, randomly scattered and their waves remain coherent over indefinitely long distances. We saw in Chapter 9 [eqn. (9.12a)] that each pair may be represented by a wavefunction which we can write as

$$\Phi_p = \Phi e^{i(\mathbf{P}\cdot\mathbf{r})/\hbar} \tag{11.1}$$

where $\mathbf{P}$ is the net momentum of the pair whose centre of mass is at $\mathbf{r}$. The term $e^{i(\mathbf{P}\cdot\mathbf{r})/\hbar}$ has the form of a travelling wave and describes the motion of the centre of mass of the pair.

It was shown in § 9.3.7 that, if the current density is uniform, all the electron-pairs in a superconductor have the same momentum and therefore have waves of the same wavelength. The superposition of a number of coherent waves of equal wavelength simply results in another wave with the same wavelength, so all the electron-pairs in a superconductor can together be described by a single wave of a form similar to (11.1), i.e.

$$\Psi_p = \Psi e^{i(\mathbf{P}\cdot\mathbf{r})/\hbar} \tag{11.2}$$

where $|\Psi_p|^2$ is the density of electron-pairs and $\mathbf{P}$ is the momentum per pair. We shall call this wave, describing the motion of *all* the electron-pairs, the *electron-pair wave*. The electron-pair wave retains its phase coherence over indefinitely long distances. This chapter is concerned with some of the phenomena which occur as the result of this long-range coherence. These phenomena are analogous to the interference and diffraction effects observed with ordinary electromagnetic waves, and, because they are manifestations on a macroscopic scale of quantum behaviour, the phenomena are often referred to under the collective title of "quantum interference".

11.1.1. Phase of the electron-pair wave

It is important to understand what the coherence of the electron-pair wave implies. Coherence of a wave travelling through a region means that if we know the phase and amplitude at any point we can, from a knowledge of the wavelength and frequency, calculate the phase and amplitude at any other point. In other words, because the wave travels undisturbed, the amplitudes and relative phases at all points in the region are uniquely related by the wave equation.

Let us rewrite the electron-pair wave (11.2) in the form†

$$\Psi_P = \Psi \sin 2\pi\left(\frac{x}{\lambda} - vt\right)$$

(for simplicity we consider a one-dimensional wave), and suppose that the wave frequency v is related to the total kinetic energy E of a Cooper pair by the familiar relation $E = hv$, and that the wavelength λ is related to the momentum P of the pair's centre of mass by the de Broglie relation $\lambda P = h$.

Consider a length of superconductor joining two points X and Y. If no current is flowing between X and Y, the momentum P of the electron pairs is zero and the wavelength λ is infinite. Consequently, the phase of an electron-pair wave is the same at X as at Y.

Suppose now that a resistanceless current flows from X to Y. The electron pairs now have a momentum P and the electron-pair wave a finite wavelength $\lambda = h/P$. There will therefore be a phase difference $(\Delta\phi)_{XY}$ between the points X and Y and this phase difference remains constant in time. The phase difference between two points past which a plane wave is travelling is:

---

† In accordance with the usual convention, the time-dependent oscillatory factor $e^{-i2\pi vt}$ has been omitted from the expression for the wavefunction in (11.1).

$$(\Delta\phi)_{XY} = \phi_X - \phi_Y = 2\pi \int_X^Y \frac{\hat{\mathbf{x}}}{\lambda} \cdot d\mathbf{l},$$

where $\hat{\mathbf{x}}$ is a unit vector in the direction of the wave propagation, and $d\mathbf{l}$ is an element of a line joining X to Y. (Though $\hat{\mathbf{x}}$ is, by definition, parallel to the direction of propagation, the wave itself need not propagate in a straight line. Later in this chapter we shall need to consider waves travelling round a closed circular path.) Now for the electron-pair wave $\lambda = h/P$ and the pair momentum is $P = 2mv$, where v is the velocity of the pairs due to the current. The relation of v to the supercurrent density $\mathcal{J}_s$ is $\mathcal{J}_s = \frac{1}{2}n_s \cdot 2e \cdot v$, where n_s is the density of superelectrons, and $\frac{1}{2}n_s$ is the density of electron pairs. The wavelength is therefore

$$\lambda = \frac{hn_s e}{2m\mathcal{J}_s},$$

so the phase difference between X and Y due to the current is

$$(\Delta\phi)_{XY} = \frac{4\pi m}{hn_s e} \int_X^Y \mathbf{J}_s \cdot d\mathbf{l}, \tag{11.3}$$

since $\hat{\mathbf{x}}$ is necessarily parallel to $\mathbf{J}_s$.

Because the electron-pair wave retains its coherence throughout a superconductor it should be able to produce long-range interference phenomena, and it should be possible to observe effects analogous to familiar optical interference, e.g. Fraunhofer diffraction and diffraction by a grating. As we shall see, such effects in superconductors can be demonstrated experimentally, and such demonstrations provide a convincing justification for treating superelectrons as waves with long-range coherence.

11.1.2. Effect of a magnetic field

The phase of the electron-pair wave may be strongly affected by the presence of an applied magnetic field. We saw in § 8.5 that in the presence of a magnetic field the momentum $\mathbf{p}$ of particles with charge q takes the form $m\mathbf{v} + q\mathbf{A}$, where $\mathbf{A}$ is the magnetic vector potential defined by curl $\mathbf{A} = \mathbf{B}$. In the case of electron pairs, $\mathbf{P} = 2m\mathbf{v} + 2e\mathbf{A}$. We can still derive the wavelength λ from the momentum P by $\lambda = h/P$, and so, by the same argument which led to (11.3) we find that in the presence of a magnetic field, the phase difference between two points X and Y is

$$(\Delta\phi)_{XY} = \frac{4\pi m}{hn_s e}\int_X^Y \mathbf{J}_s \cdot d\mathbf{l} + \frac{4\pi e}{h}\int_X^Y \mathbf{A} \cdot d\mathbf{l}. \qquad (11.4)$$

We interpret the first term on the right-hand side as being the phase difference due to the current, and the second term as an additional phase difference due to the magnetic field. We can therefore write

$$(\Delta\phi)_{XY} = [(\Delta\phi)_{XY}]_I + [(\Delta\phi)_{XY}]_B,$$

where $[(\Delta\phi)_{XY}]_I$ is the phase difference due to any current, and $[(\Delta\phi)_{XY}]_B$ is the phase difference due to any magnetic field. We have, therefore,

$$[(\Delta\phi)_{XY}]_I = \frac{4\pi m}{hn_s e}\int_X^Y \mathbf{J}_s \cdot d\mathbf{l} \qquad (11.5)$$

and

$$[(\Delta\phi)_{XY}]_B = \frac{4\pi e}{h}\int_X^Y \mathbf{A} \cdot d\mathbf{l}. \qquad (11.6)$$

The phase difference which can be produced by the presence of a magnetic field plays a very important part in the phenomena which are now to be described.

11.2. The Fluxoid

We now consider the case of a supercurrent circulating around a closed path. The results of this section will be of importance in a later chapter.

Figure 11.1 shows a superconductor enclosing a non-superconducting region N. Suppose that in N there is a flux density $\mathbf{B}$ due to supercurrents flowing around it (see § 2.3.1). Such a situation might arise where N is a hole through the material or, in the case of a solid piece of superconductor, where the magnetic field generated by the encircling current maintains the region N in the normal state.

Consider a closed path, such as the dotted curve in Fig. 11.1, which encircles the normal region. There will be a phase difference of the electron-pair wave between any two points on this curve, due both to the presence of the magnetic field and to the circulating current. As we have seen, the phase difference between the points X and Y is given by (11.4):

$$(\Delta\phi)_{XY} = \frac{4\pi m}{hn_s e}\int_X^Y \mathbf{J}_s \cdot d\mathbf{l} + \frac{4\pi e}{h}\int_X^Y \mathbf{A} \cdot d\mathbf{l}.$$

COHERENCE OF THE ELECTRON-PAIR WAVE; QUANTUM INTERFERENCE 157

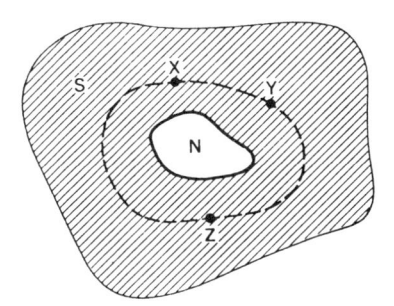

FIG. 11.1. Superconductor enclosing a non-superconducting region.

Now consider the phase change occurring around a closed path, say *XYZX*. The total phase change will be

$$\Delta\phi = \frac{4\pi m}{hn_s e}\oint \mathbf{J}_s \cdot d\mathbf{l} + \frac{4\pi e}{h}\oint \mathbf{A} \cdot d\mathbf{l}.$$

Now by Stokes' theorem, $\oint \mathbf{A} \cdot d\mathbf{l} = \iint_S \text{curl } \mathbf{A} \cdot d\mathbf{S}$, where $d\mathbf{S}$ is an element of area, and furthermore curl $\mathbf{A} = \mathbf{B}$, so we can rewrite $\oint \mathbf{A} \cdot d\mathbf{l}$ as $\iint_S \mathbf{B} \cdot d\mathbf{S}$, where S is the area enclosed by *XYZX*, and the phase change round the closed path may be rewritten

$$\Delta\phi = \frac{4\pi m}{hn_s e}\oint \mathbf{J}_s \cdot d\mathbf{l} + \frac{4\pi e}{h}\iint_S \mathbf{B} \cdot d\mathbf{S}.$$

What follows in this chapter depends on the realization that, if the superelectrons are to be represented by a wave, the wave at any point can, at any instant, have only one value of phase and amplitude. Consequently, the phase change $\Delta\phi$ around a closed path must equal $2\pi n$ where n is any integer. We shall refer to this as the "phase condition" or "quantum condition".† We have therefore

$$\frac{4\pi m}{hn_s e}\oint \mathbf{J}_s \cdot d\mathbf{l} + \frac{4\pi e}{h}\iint_S \mathbf{B} \cdot d\mathbf{S} = 2\pi n \qquad (11.7)$$

which can be rewritten

$$\frac{m}{n_s e^2}\oint \mathbf{J}_s \cdot d\mathbf{l} + \iint_S \mathbf{B} \cdot d\mathbf{S} = n\frac{h}{2e}.$$

† This condition is also responsible for the quantized electron orbits in the Bohr atom.

F. and H. London named the quantity on the left-hand side of this equation the *fluxoid* enclosed by the curve $XYZX$. It is given the symbol Φ' to distinguish it from the flux Φ:

$$\Phi' = \frac{m}{n_s e^2} \oint \mathbf{J}_s \cdot d\mathbf{l} + \iint_S \mathbf{B} \cdot d\mathbf{S}, \tag{11.8}$$

and we have shown that, because of the single-valuedness of the wave, the fluxoid can only exist in integral multiples of the unit $h/2e$:

$$\Phi' = n \frac{h}{2e}. \tag{11.9}$$

Note that it is sometimes convenient to write the fluxoid in the equivalent form

$$\Phi' = \frac{m}{n_s e^2} \oint \mathbf{J}_s \cdot d\mathbf{l} + \oint \mathbf{A} \cdot d\mathbf{l}. \tag{11.10}$$

The fluxoid within a closed curve is closely related to, though not identical with, the magnetic flux within the curve. The first term on the right-hand side of (11.8) contains the line integral of the current density around the curve $XYZX$, but because the penetration depth is small, nearly all the circulating current will in fact be concentrated very close to the boundary of the normal region N, and this term will be negligible unless we are considering a curve of which a considerable part lies very close to N. The second term on the right-hand side of (11.8) is just the magnetic flux contained within N and the penetration depth around it. So unless we are considering a closed curve which lies very close to the boundary of N, the fluxoid which a curve encloses is practically the same as the flux it encloses.

We see, therefore, that any flux (strictly fluxoid) contained within a superconductor should only exist as multiples of a quantum, the *fluxon*, Φ_0, given by

$$\Phi_0 = h/2e \tag{11.11}$$

$$= 2 \cdot 07 \times 10^{-15} \text{ weber}.$$

It can be seen that this predicted value of the fluxon is extremely small. The value of the fluxon has been measured experimentally and the fact that its value is found to be, as predicted, Planck's constant divided by *twice* the electronic charge, is strong evidence that the supercurrent is carried by *pairs* of electrons.

What is measured in experiments is the flux, not the fluxoid, but as we have seen, in most cases these are indistinguishable, and the expression "quantized flux" is often used instead of the strictly accurate "quantized fluxoid". Measurements have, for example, been made of the magnetic moment of a long, hollow cylinder of superconductor which was repeatedly cooled below its transition temperature in very weak axial magnetic fields.† The thickness of the wall was small compared to the diameter of the central hole (though large compared to the penetration depth) and consequently the magnetic moment of the cylinder was proportional to the flux trapped in the hole. The measurements showed that the cylinder could only trap an amount of flux corresponding to an integral number of fluxons.

Suppose a ring or hollow cylinder becomes superconducting while in an applied magnetic field. We have seen that once the material has become superconducting the flux threading the hole can only be an integral number of fluxons $n\Phi_0$. However, in general, the strength of the applied field will not be such that the flux from it which threads the hole is an exact number of fluxons, though the difference will be small because the fluxon is small. When the ring becomes superconducting the strength of the circulating current which arises on the inner surface to maintain flux in the hole is just that which, together with the applied field, produces the nearest integral number of fluxons. For example, if the applied field happened to produce a flux $(n + \frac{1}{4})\Phi_0$ in the hole, then in the superconducting state the flux in the hole will become $n\Phi_0$. If, however, the applied field strength had been such as to produce a flux $(n + \frac{3}{4})\Phi_0$ in the hole, the magnitude of the circulating current will be such as to produce a flux of $(n + 1)\Phi_0$ (for an explanation of this see § 11.4).

Flux quantization is a special property of superconductors and does not occur in normal metals for the reason discussed in § 11.1. We shall see that fluxons play an important role in the properties of type-II superconductors which are discussed in the second part of this book.

11.2.1. Fluxoid within a superconducting metal

In the previous section we have been considering a closed curve which surrounds a non-superconducting region, e.g. a hole. Magnetic flux can thread this hole accompanied by a current flowing round the hole

---

† For example, Doll and Näbauer, *Phys. Rev. Letters* **7**, 51 (1961). Deaver and Fairbank, *ibid.* **7**, 43 (1961).

(§ 2.3.1). The fluxoid enclosed within the curve will be an integral number n of fluxons, but the value of n will be zero if no flux threads the hole. If, however, we consider a closed curve which does not encircle a non-superconducting region, so that the area enclosed by the curve is entirely superconducting, then n is always zero. This will be so even if the curve passes close to a boundary of the superconductor where, within the penetration depth, neither the flux density $\mathbf{B}$ nor the current density $\mathbf{J}_s$ need be zero. The magnitudes and directions of $\mathbf{B}$ and $\mathbf{J}_s$ are everywhere such that, when integrated around the curve, the two terms contributing to the fluxoid [right-hand of (11.8)] cancel to give zero.

The statement that the fluxoid within any closed curve not surrounding a non-superconducting region is zero is a more precise way of expressing the perfect diamagnetism of a superconducting material, since it is valid everywhere in the metal, including the penetration depth.

11.3. Weak Links

11.3.1. Josephson tunnelling

Consider two superconductors P and Q completely isolated from each other; the phase of the electron-pair wave in P will be unrelated to the phase of the wave in Q. Suppose now that the gap separating the two pieces is gradually reduced to zero. When the separation becomes very small, electron pairs can tunnel across the gap and the electron-pair waves in P and Q tend to become coupled together. As P and Q approach each other there is an increase in the interaction between their electrons due to the tunnelling, and the phases of the waves in the two pieces become progressively more tightly locked together. Eventually when P and Q come into contact, they form one single piece of metal and there must, under a given set of external conditions, be a definite relation between the phases throughout P and Q.

Before contact is made, the interaction occurs as a result of the spread of the electron-pair wave through the gap, i.e. a tunnelling of electron pairs from P and Q and vice versa. Tunnelling of an electron *pair* means that the two electrons maintain their momentum pairing after crossing the gap. We have already met this type of tunnelling in the previous chapter, where it was referred to as "Josephson tunnelling". If the gap is thin the tunnelling across it of electron pairs is relatively probable and so an appreciable resistanceless current can flow through it. A gap,

COHERENCE OF THE ELECTRON-PAIR WAVE; QUANTUM INTERFERENCE 161

however, has a critical current as does an ordinary superconductor. When resistanceless current flows across such a gap by electron-pair tunnelling there is a phase difference between the electron-pair waves on each side of the gap. It can be shown that if i_s is the supercurrent crossing the gap from P to Q

$$i_s = i_c \sin(\phi_Q - \phi_P) \tag{11.12}$$

where ϕ_Q and ϕ_P are the phases on each side of the gap. The maximum value of the supercurrent occurs when there is a phase difference of $\pi/2$ across the gap, and i_s then equals the critical current i_c of the gap.

It is important to realize that, because the phase difference in (11.12) is not restricted to the range 0 to $\pi/2$, the phase difference is not uniquely determined for a given value of i_s but may take one of two alternative values, $\Delta\phi$ or $\pi - \Delta\phi$.†

Equation (11.12) relates the electron-pair current tunnelling across a Josephson junction to the phase difference of the electron-pair wave on the two sides. We saw in §§ 1.4 and 3.1 that if the supercurrent in a superconductor is varying with time, a voltage is developed across the superconductor. The same is true of a Josephson tunnelling junction. If the current through the junction is varying with time, the phase difference $\Delta\phi$ across it must also be changing with time, and it can be shown that a voltage V is developed across it, related to the rate of change of the phase difference by

$$2eV = \hbar \frac{d}{dt} \Delta\phi. \tag{11.13}$$

It can be seen that this equation is consistent with equation (10.1). Equations (11.12) and (11.13) show how the current through a Josephson tunnelling junction and any voltage which may be developed across it are related to the phase difference of the electron-pair waves on the two sides:

$$i_s = i_c \sin \Delta\phi \tag{11.14.i}$$

$$2eV = \hbar \frac{d}{dt} \Delta\phi \tag{11.14.ii}$$

† The phase difference may also take on values $2\pi n + \Delta\phi$ or $2\pi n + (\pi - \Delta\phi)$ but these are not physically distinguishable from $\Delta\phi$ and $\pi - \Delta\phi$.

These two equations are the basic equations of a Josephson tunnelling junction, and from them we can derive nearly all its properties. They follow from the coherent nature of the electron-pair wave in a superconductor. We do not derive them in this book but a fairly simple derivation is given by Feynman.†

11.3.2. Pendulum analogue

We shall now show that there is a close analogy between a Josephson tunnelling junction and a simple pendulum. This analogy is very useful because the mathematics of Josephson junctions can be rather complicated, and the results difficult to interpret, but it is often possible to visualize intuitively how a simple pendulum will behave and draw conclusions through the analogy about the behaviour of a Josephson junction. Furthermore, even if one cannot predict how a pendulum will behave under a certain set of circumstances, it is relatively easy to perform experiments on a pendulum and obtain results which can be transferred to a Josephson junction.

To establish the analogy, consider the various mechanisms by which current may flow through a Josephson junction. First there is the electron-pair tunnelling current i_s. We consider the general case in which this current varies with time producing, as pointed out on p. 161, a voltage V across the junction. Because there is a voltage across the junction, a current of normal (unpaired) electrons will also flow across the junction by the normal tunnelling process. This process is resistive and can be represented by a resistance R across the junction [Fig. 11.2(a)]. Because the junction consists of two metal surfaces very close together, there is also a capacitance C across the junction. The three parallel branches in Fig. 11.2(a) therefore form the equivalent circuit of a Josephson tunnelling junction. If we pass a current I through the junction from an external source, this current must equal the sum of the currents flowing down each of the three branches of the equivalent circuit:

$$I = C\frac{dV}{dt} + \frac{V}{R} + i_c \sin \Delta\phi.$$

To find how the total current is related to the phase difference $\Delta\phi$ across the junction, we use eqn. (11.14.ii) to replace V by $(\hbar/2e)(d/dt)\Delta\phi$ with the result that

† R. P. Feynman, *Lectures on Physics*, Vol. III, pp. 21–16 (Addison-Wesley).

COHERENCE OF THE ELECTRON-PAIR WAVE; QUANTUM INTERFERENCE 163

(a) Equivalent circuit of Josephson junction

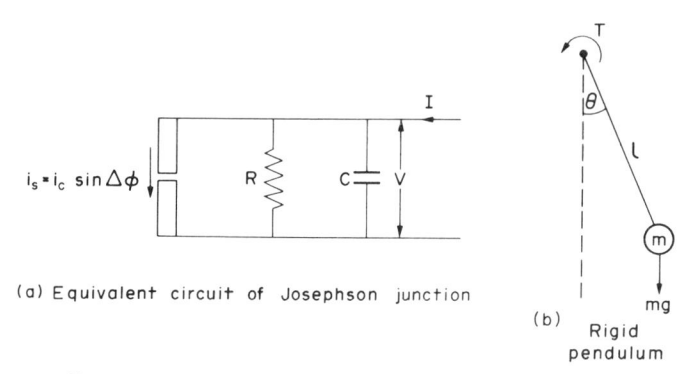

(b) Rigid pendulum

FIG. 11.2. Pendulum analogue of a Josephson tunnelling junction.

$$I = \frac{C\hbar}{2e}\frac{d^2}{dt^2}\Delta\phi + \frac{\hbar}{2eR}\frac{d}{dt}\Delta\phi + i_c \sin \Delta\phi, \quad (11.15)$$

which is the equation relating the total current through a Josephson tunnelling junction to the phase difference of the electron-pair waves on each side.

Let us now consider the simple rigid pendulum of Fig. 11.2(b) which consists of a light stiff rod of length l with a bob of mass m at its lower end. The pendulum can rotate freely about the pivot P. If an external torque T is applied the pendulum will swing out of the vertical. Let the angle of deflection at any instant be θ. By analogy with Newton's second law, torque produces a rate of change of angular momentum, so if M is the moment of inertia of the pendulum about P,

$$M\frac{d^2\theta}{dt^2} = \text{total torque}. \quad (11.16)$$

Now the total torque acting on the pendulum consists of several parts: there is the applied torque T which deflects the pendulum; this is opposed by the weight of the bob which exerts a torque $-mgl \sin \theta$ and, if the pendulum is not in a vacuum and rotates with velocity $d\theta/dt$, the viscosity of the air will exert an opposing torque $-\eta d\theta/dt$. So (11.16) can be written

$$M\frac{d^2\theta}{dt^2} = T - mgl \sin \theta - \eta \frac{d\theta}{dt}.$$

Rearrangement to bring all terms in θ together gives

$$T = M\frac{d^2\theta}{dt^2} + \eta \frac{d\theta}{dt} + mgl \sin \theta. \quad (11.17)$$

This is the equation of rotational motion of a simple pendulum. We can see that this equation has exactly the same form as equation (11.15) which relates the total current through a Josephson junction to the phase difference of the electron-pair wave on each side. A rigid pendulum is therefore an analogue of a Josephson tunnelling junction. Comparison of the two equations term by term,

$$\text{(junction)}\ I = \frac{C\hbar}{2e}\frac{d^2}{dt^2}\Delta\phi + \frac{1}{R}\frac{\hbar}{2e}\frac{d}{dt}\Delta\phi + i_c \sin\Delta\phi, \quad (11.18)$$

$$\text{(pendulum)}\ T = M\frac{d^2}{dt^2}\theta + \eta\frac{d}{dt}\theta + mgl \sin\theta \quad (11.19)$$

gives the following correspondence between the mechanics of the pendulum and the electrical properties of the junction:

Junction	*Pendulum*
Phase difference, $\Delta\phi$	Deflection, θ
Total current across junction, I	Applied torque, T
Capacitance, C	Moment of inertia, M
Normal tunnelling conductance, $1/R$	Viscous damping, η
Electron-pair tunnelling current, $i_s = i_c \sin\Delta\phi$	Horizontal displacement of bob, $x = l \sin\theta$
Voltage across junction, $V = \frac{\hbar}{2e}\frac{d}{dt}\Delta\phi$	Angular velocity, $\omega = \frac{d\theta}{dt}$

These analogues are summarized in Fig. 11.3.

As pointed out earlier, this analogue is very useful because by visualizing or experimenting on the motion of a pendulum we can deduce the electrical behaviour of a Josephson tunnelling junction. As an example, we now investigate how the voltage across a junction is related to the current through it. We represent a gradual increase of current through the junction by a gradual increase of torque applied to the pendulum. We can imagine the torque to be provided by weights hanging from a drum attached to the pivot of the pendulum, as in Fig. 11.3. When a small torque is applied (i.e. a small current passed through the junction) the pendulum finally settles down at a constant angle of deflection θ. There is then no angular velocity, so this implies that there is no voltage across a junction when a small current is passing through it, i.e. the junction is superconducting. If the torque is gradually increased, the pendulum deflects to a greater but steady angle, i.e. we can pass more

COHERENCE OF THE ELECTRON-PAIR WAVE; QUANTUM INTERFERENCE 165

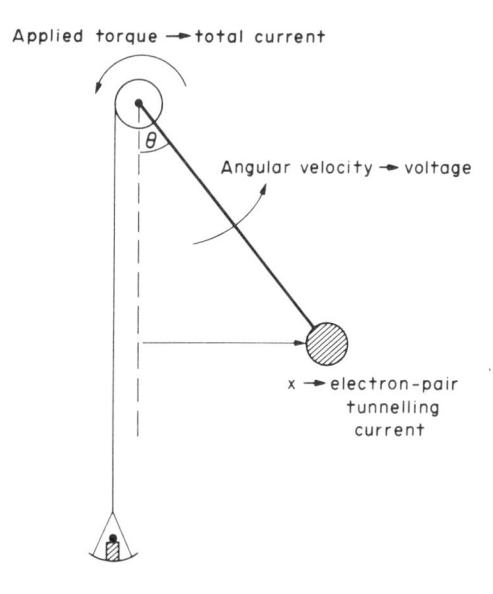

FIG. 11.3. Analogy between pendulum and Josephson tunnelling junction.

current through a junction without any voltage appearing. There is, however, a maximum torque which can be applied to the pendulum and which still produces a stationary deflection. This is the torque which deflects the pendulum through a right angle so that it is horizontal. If we apply any greater torque the pendulum rises, accelerates upwards, passes through the vertically "up" position, and thereafter continues to rotate continuously around its axis so long as the torque continues to be applied. We see, therefore, that if more than a certain "critical torque" is applied the pendulum cannot remain at rest but rotates continuously. Because angular velocity is the analogue of voltage across a junction and the angular velocity is always in the same direction, this unstable behaviour of the pendulum shows that a d.c. voltage will appear across a junction if the current passed through it exceeds a critical value, i.e. a junction has a critical current.

11.3.3. a.c. Josephson effect

Let us consider further what can be deduced from the pendulum analogue of a Josephson tunnelling junction. Suppose that a pendulum is rotating continuously because a torque greater than the critical torque is being applied to it. As the pendulum rotates, the horizontal deflection x of the bob (Fig. 11.3) oscillates, changing from right to left and back again. We have seen that the horizontal deflection of the weight corresponds to the electron-pair current tunnelling across the junction. So the analogue tells us that when there is a *d.c.* voltage across a Josephson tunnelling junction an *a.c.* electron-pair current tunnels back and forth across it. This is the *a.c. Josephson effect* mentioned in § 10.6. What is the frequency of the a.c. tunnelling current? To answer this we return to our pendulum analogue. When a constant torque T, greater than the critical torque, is applied to a pendulum, the pendulum rotates, and its rotation accelerates until the energy lost due to viscous damping during each rotation equals the work done on the pendulum by the constant applied torque. When this state is reached the period of each revolution is the same. The angular velocity, however, is not constant but varies during each revolution (on the half-cycle during which the bob is rising the rotation decelerates because gravity opposes the applied torque, but on the following half-cycle the bob is accelerated by gravity as it falls). The frequency of the rotation is $(1/2\pi)\langle d\theta/dt \rangle$ where $\langle d\theta/dt \rangle$ is the time average of the angular velocity over one cycle. We deduce, therefore, that the electron-pair tunnelling current, which is the analogue of the horizontal displacement of the pendulum's bob, oscillates back and forth across the junction with frequency ν equal to $(1/2\pi)\langle (d/dt)\Delta\phi \rangle$. Now we have seen that when the phase difference across a junction is varying at a rate $(d/dt)\Delta\phi$ a voltage V appears across the junction whose instantaneous value is $(\hbar/2e)(d/dt)\Delta\phi$ [eqn. (11.14.ii)]. By analogy with the pendulum $(d/dt)\Delta\phi$ always has the same sign but is not constant, so that the voltage V contains both d.c. and a.c. components, i.e. it is a d.c. voltage with an a.c. ripple on it. The average value $\langle (d/dt)\Delta\phi \rangle$ of the rate of change of phase across the junction is given by

$$\left\langle \frac{d}{dt}\Delta\phi \right\rangle = \frac{2e}{\hbar}\langle V \rangle = \frac{2e}{\hbar}V_{\mathrm{dc}},$$

because the d.c. component is the time average of the voltage. Consequently the frequency of oscillation is related to the d.c. voltage across the junction by

$$\nu = \frac{1}{2\pi}\left\langle \frac{d}{dt}\Delta\phi \right\rangle = \frac{2e}{h}V_{\mathrm{dc}}.$$

The a.c. Josephson effect can be observed through the a.c. component of the voltage which appears across the junction and which can be detected by suitable instrumentation. This alternating component of the voltage causes emission of radiation with frequency ν given by $h\nu =$

$2eV_{dc}$. This is in agreement with eqn. (10.1) derived from consideration of the energy balance of the tunnelling electron pairs.

We may sum up the situation as follows: when a steady direct current, greater than the critical current, is fed through a Josephson junction from a constant current source a d.c. voltage V_{dc} appears across the junction. The electron-pair tunnelling current has, however, a large oscillatory component which tunnels back and forth with frequency $(2e/h)V_{dc}$ across the junction. This a.c. Josephson tunnelling can be observed as a voltage ripple of frequency $(2e/h)V_{dc}$ superimposed on the d.c. voltage.

The a.c. Josephson effect forms the basis of a number of useful devices. An oscillating Josephson junction can be used as a tunable oscillator whose frequency ν is given by $\nu = 2eV/h$, where V is the d.c. voltage across the junction. The power output is very small but its usefulness lies in the fact that the frequency can be very accurately controlled simply by adjusting the voltage across the junction. Josephson junctions are also used as very sensitive detectors of radiation; because an alternating electric field acting on a biased oscillating junction can drive a d.c. current through an external load.

11.3.4. Coupling energy

We have seen that the supercurrent i_s tunnelling through a Josephson junction is sinusoidally related to the phase difference of the electron-pair waves on the two sides [eqns. (11.14.i) and (11.14.ii)]. If, therefore, a certain current, i_1 say, is passed through the junction there are two possible phase differences across it, $\Delta\phi_1$ and $\pi - \Delta\phi_1$ (Fig. 11.4). We naturally ask which of these two possible phase differences will in fact occur. The answer is determined by the fact that, as we shall now show, the energy of a Josephson tunnelling junction depends on the phase difference across it. Consider a junction through which a constant current i_s is being passed, the current having been raised from zero to this value over a time t. During the time the current is increasing there must be a rate of change of current di/dt, but eqn. (11.14.ii) tells us that if the current through a junction is changing a voltage will appear across it, so during the time t that the current is being established there is a voltage V across the junction, and power iV is being delivered to it. So work, $W = \int_0^t iV dt$, is performed in setting up the current and the consequent phase difference across it, and W must be the increase in energy

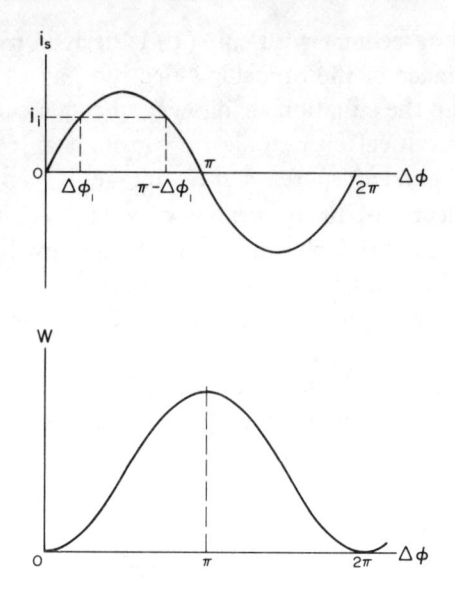

FIG. 11.4. Josephson junction; relationship of tunnelling current i_s and energy W to phase difference $\Delta\phi$.

of the junction due to the passage of a current through it. Equations (11.14.i) and (11.14.ii) relate the voltage and current to the phase difference, and substitution of these gives the increase in junction energy as

$$W = \frac{i_c \hbar}{2e} \int_0^t \sin \Delta\phi \cdot \frac{d}{dt} \Delta\phi \cdot dt,$$

$$W = \frac{i_c \hbar}{2e} [1 - \cos \Delta\phi]. \quad (11.20)$$

This relationship is shown in Fig. 11.4. We see that the energy W increases with $\Delta\phi$ for phase differences up to π and so, if a current i_1 is passed through the junction it is energetically favourable for the smaller phase difference $\Delta\phi_1$ to establish itself across the junction rather than the larger difference $\pi - \Delta\phi_1$. When there is no current through the junction, the phase difference across it will be zero not π; in terms of our pendulum analogue this is equivalent to saying that, for energetic reasons, if no torque is applied the pendulum hangs vertically downwards not vertically upwards.

11.3.5. Weak-links

Josephson junctions are a special case of a more general kind of weak contact between two superconductors. These other forms have properties similar to the Josephson tunnelling junctions we have just described. A resistanceless current can be passed through a very narrow constriction or a point contact between two superconductors (Fig. 11.5) while a phase difference is maintained between the superconductors on each side. We shall include all these configurations under the general name of "weak-links". A weak-link is a region which has a much lower critical current than the superconductors it joins and into which an applied magnetic field can penetrate.

(a) Tunnelling junction

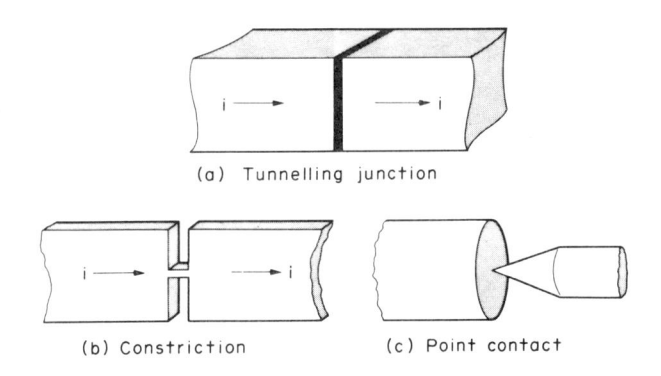

(b) Constriction   (c) Point contact

FIG. 11.5. Superconducting weak-links.

For all weak-links, including Josephson tunnelling junctions, the resistanceless current flowing through the link increases as the phase difference across the link increases, and the critical current is the current at which the phase difference reaches $\pi/2$. However, the current is not in general proportional to the *sine* of the phase difference; in this respect the Josephson junction is a special case.

Though the behaviour of weak-links which are not Josephson tunnelling junctions may differ in detail from that of a Josephson tunnelling junction their general behaviour is very similar, and the effects described in this chapter for Josephson tunnelling junctions also occur for the other forms of weak-link.

Weak-links in the form of constrictions or point contacts are usually better than tunnelling junctions as radiation sources or detectors because radiation can more easily be coupled into or out of them.

11.4. Superconducting Quantum Interference Device (SQUID)

We now consider a superconducting device whose properties well illustrate the coherence of the electron-pair wave throughout a superconductor and the effect of weak-links on this coherent wave. The device is called a *superconducting quantum interference device*, usually abbreviated to "SQUID". SQUIDs can be used to detect and measure extremely weak magnetic fields and are the basic elements of a range of very sensitive and useful instruments. The basic element of a SQUID is a ring of superconducting metal containing one or more weak-links. We shall consider here the particular form which includes two weak-links, as shown in Fig. 11.6. The ring of superconducting material has two similar weak-links X and W whose critical current i_c is very much less than the

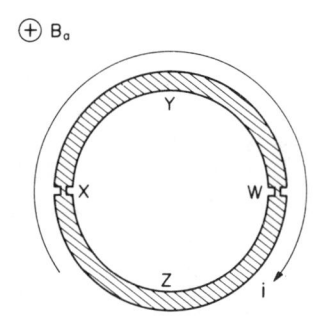

FIG. 11.6. Superconducting ring with two weak-links.

critical current of the rest of the ring. Because supercurrents flowing round such circuits cannot exceed the critical current of the weak-links, resistanceless currents must be very much less than the critical currents of the superconductors joined by the weak-links. Because of the low current density in the superconductors, the momentum of the Cooper pairs in them is small and the corresponding electron-pair waves have a very long wavelength. There is, consequently, very little phase difference between different parts of any one superconductor, and we shall assume that this difference is negligible, so that, if there is no applied magnetic field, the phase is the same throughout any one piece of superconductor which does not include a weak link. As a result, a phase difference appears across the superconductors only when a magnetic field is applied.

We now examine the effect of the application of a magnetic field to a SQUID. In studying the behaviour of this circuit, we shall apply two of the results discussed in the previous sections. First, that if a magnetic

COHERENCE OF THE ELECTRON-PAIR WAVE; QUANTUM INTERFERENCE   171

field is applied perpendicular to the plane of the ring, it produces a difference in phase of the electron-pair wave along the path XYW and along the path WZX (Fig. 11.6); we assume that the weak-links X and W are so short that the phase difference across them produced by the applied magnetic field is negligible. Second, that if a small current flows round the ring, it produces a phase difference across the weak-links, but negligible phase difference across the thick segments XYW and WZX.

Suppose that the ring has been cooled below its transition temperature in the absence of an applied magnetic field so that there is no magnetic flux threading the hole. Now a magnetic field of gradually increasing flux density B_a is applied perpendicular to the plane of the ring. If the ring had no weak-links the application of the magnetic field would induce a current i (Fig. 11.6) which would circulate to cancel the flux in the hole. We shall use the convention that the clockwise direction in Fig. 11.6 has a positive sign. For the cancellation to be complete the magnitude $|i|$ of the circulating current would have to be such that

$$L|i| = \Phi_a \qquad (11.21)$$

where L is the inductance of the ring and $\Phi_a = \mathscr{A} B_a$, $\mathscr{A}$ being the area of the hole enclosed by the ring.

The presence of the weak-links has two effects:

(i) They have a very small critical current i_c, so any circulating current must be less than this value. Consequently, unless B_a is very small, not enough current can circulate to cancel the flux in the hole, and this flux can no longer be maintained at zero.

(ii) Even though any circulating current is limited by the weak-links to a maximum value of i_c, it can nevertheless introduce a significant phase change across each of them.

The particular form of SQUID we are about to consider is constructed so that both the critical current i_c of the weak-links and the inductance L of the ring are extremely small, so that

$$i_c L \ll \Phi_0. \qquad (11.22)$$

Now iL is the flux generated in the hole by a circulating current, so the presence of the weak-links means that an induced circulating current cannot generate even one fluxon, and the net flux Φ in the hole is scarcely different from the flux Φ_a of the applied magnetic field;

$$\Phi \simeq \Phi_a.$$

The presence of the weak-links also means that the flux within the hole is no longer necessarily an integral number of fluxons.† However, the quantum condition that the total phase change around any closed path must equal $n2\pi$ can still be satisfied because the circulating current, though unable to generate even one fluxon, can produce considerable phase differences across each weak-link.

Equation (11.6) tells us that the application of a magnetic field produces a phase difference $\Delta\phi(B)$ around the ring

$$\Delta\phi(B) = \frac{4\pi e}{h}\oint \mathbf{A} \cdot d\mathbf{l}.$$

Now $\oint \mathbf{A} \cdot d\mathbf{l}$ is the flux Φ_a produced in the ring by the applied magnetic field (we have seen that the flux Li produced by the circulating current is negligible, eqn. (11.22)), and $h/2e$ equals the fluxon Φ_0, so

$$\Delta\phi(B) = 2\pi \frac{\Phi_a}{\Phi_0}. \tag{11.23}$$

So, when a magnetic field is applied to the superconducting ring, its effect on the electron-pair wave is to produce a phase change around the ring which is proportional to the flux of the applied field, as shown in Fig. 11.7(a). If, for example, the flux density B_1 of the applied field is such that the flux threading the ring is Φ_1 (Fig. 11.7(a)), the phase change it produces around the ring is $\Delta\phi(B_1)$. In general the strength of the applied magnetic field will not be such as to produce an integral number of fluxons in the hole and the resulting phase change $\Delta\phi(B_1)$ will not equal a multiple of 2π. However, the phase change round any closed superconducting circuit *must* equal an integral multiple of 2π; a current i therefore circulates round the ring of such a strength that the additional phase different $2\Delta\phi(i)$ it produces the two weak-links brings the total phase change round the circuit to a multiple of 2π (Fig. 11.6):

$$\Delta\phi(B) + 2\Delta\phi(i) = n \cdot 2\pi. \tag{11.24}$$

But in an applied magnetic field producing a flux Φ_1 (Fig. 11.7(a)), there are two possible circulating currents which could make the total phase change round the ring an integral multiple of 2π. A current of magnitude

---

† In a different form of SQUID, which we shall not discuss, the critical current of the weak-links and the inductance are not so small. In this form of SQUID the flux in the hole is quantized and different from the flux of the applied field. This form of SQUID operates in a different manner from the one discussed in this book.

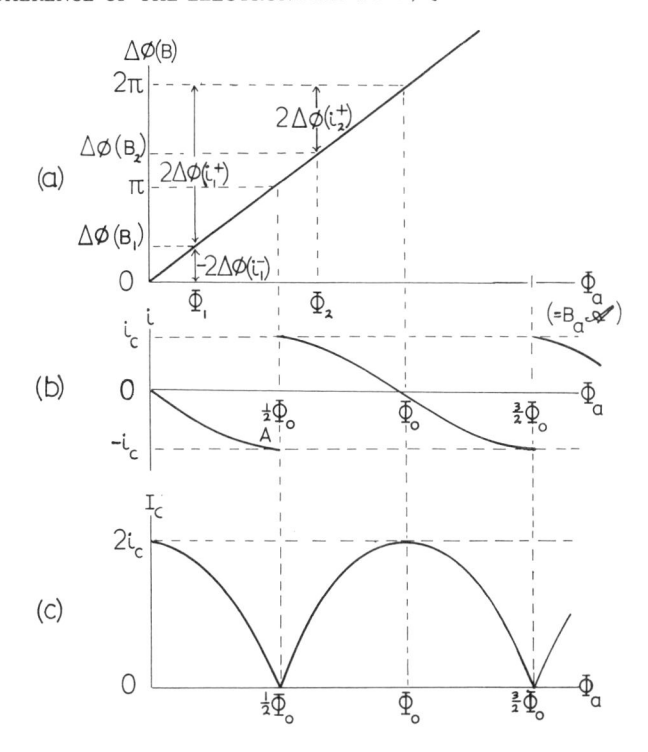

FIG. 11.7. Effect of magnetic field on SQUID with two weak-links ($Li_c \ll \Phi_0$). (a) Phase change $\Delta\phi(B_a)$ round ring due to magnetic field. (b) Circulating current i in absence of measuring current. (c) Critical measuring current I_c.

i_1^+ could circulate in the clockwise direction so that the phase difference $2\Delta\phi(i_1^+)$ (Fig. 11.7(a)) it produces across the two weak-links adds on to the phase difference $\Delta\phi(B_1)$ produced in the two halves of the ring by the magnetic field ($n = 1$), or a smaller anticlockwise current of magnitude i_1^- could circulate to produce the smaller negative phase different $-2\Delta\phi(i_1^-)$ which would exactly cancel the phase change due to the magnetic field ($n = 0$). In fact the current will circulate in the anti-clockwise direction because, as we saw in § 11.3.4, the energy of weak-links depends on the phase difference across them and, for phase differences less than π, increases as the phase difference increases. Therefore in our SQUID the smaller, anticlockwise, current is energetically favourable (because the circulating current flows through both junctions, the phase differences across them add and the phase difference across each must be less than π).

We can find how the magnitude of the circulating current depends on the strength of the applied magnetic field by means of (11.14.i) which tells us that a current i passing through a Josephson tunnelling junction† produces a phase different $\Delta\phi(i)$ given by

$$\sin \Delta\phi(i) = i/i_c. \qquad (11.25)$$

For the phase difference across the weak-links to cancel the phase difference due to the magnetic field

$$2\Delta\phi(i^-) = -\Delta\phi(B).$$

Substituting the values of $\Delta\phi(B)$ and $\Delta\phi(i^-)$ from (11.23) and (11.25) we obtain the strength of the circulating current

$$|i^-| = i_c \sin \pi \frac{\Phi_a}{\Phi_0}.$$

If the magnetic field strength is increased so that the flux through the ring approaches $\frac{1}{2}\Phi_0$ the magnitude of the anticlockwise circulating current increases as shown between O and A in Fig. 11.7(b). It can be seen that when the applied field is such that the flux through the ring equals $\frac{1}{2}\Phi_0$ the circulating current reaches the critical current i_c of the weak-links, so at this field strength the links go normal and the circulating current dies away.

Suppose the strength of the applied magnetic field is still further increased so that the flux Φ_a through the ring now has a value Φ_2 greater than $\frac{1}{2}\Phi_0$, and the phase change $\Delta\phi(B_2)$ it produces across the two halves of the ring exceeds π (Fig. 11.7(a)). Clearly it is now energetically favourable for the total phase change to be brought up to 2π by a current i_2^+ circulating in the clockwise direction to produce phase changes across the weak-links which add on to those due to the magnetic field. When Φ_a is just greater than $\frac{1}{2}\Phi_0$ the circulating current is relatively large but as the strength of the magnetic field is increased the current becomes smaller, until at Φ_0 it has fallen to zero (Fig. 11.7(b)). It can be seen that the circulating current i varies periodically with the strength of the applied field, switching from $-i_c$ to $+i_c$ when the flux through the ring is an odd multiple of $\frac{1}{2}\Phi_0$. This periodic dependence of the circulating current on the strength of an applied magnetic field is the basis of all SQUIDs.

† In this analysis we shall assume the weak-links to be Josephson tunnelling junctions. For other weak-links, such as constrictions or point contacts, where the current is periodic in $\Delta\phi$ but not sinusoidally related, a similar analysis can be carried out and leads to a qualitatively similar result.

The period of variation of the circulating current corresponds to a flux change of one fluxon, Φ_0, a very small amount of magnetic flux, so the value of the circulating current is affected by very small changes in applied magnetic field strength. If we could detect changes in the circulating current we should, therefore, be able to detect very small changes in the strength of the applied magnetic field, and our SQUID could be used as a sensitive magnetometer. We can, as we shall now see, detect changes in the circulating current by passing a current I across the ring between two side-arms Y and Z, as shown in Fig. 11.8. We shall

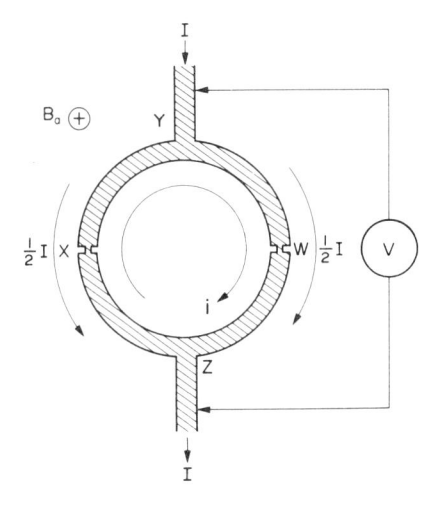

FIG. 11.8. Superconducting quantum interference device (SQUID).

call I the "measuring current". If the ring is symmetrical, I divides equally, $\frac{1}{2}I$ flowing through each weak-link; but so long as the SQUID is everywhere superconducting no voltage will be detected by the voltmeter V connected across it. However, if we increase the measuring current I enough, a voltage will be developed. The value of measuring current I at which the voltage appears is called the critical measuring current I_c of the ring. It can be seen that current $i - \frac{1}{2}I$ flows across one weak-link and a larger current $i + \frac{1}{2}I$ across the other, and it might be thought that the critical measuring current must be that which raises $i + \frac{1}{2}I$ to the critical value i_c of the weak-link through which it flows, i.e. that the critical measuring current would be $2(i_c - i)$. However, this is not necessarily true; in general it becomes impossible, as we shall now see, for the phase change of the electron-pair wave round the ring to equal an integral multiple of 2π (a necessary condition for there to be a superconducting path right round the ring) even though the current through neither weak-link has reached i_c. That is to say, a voltage appears across the ring at a measuring current I_c less than $2(i_c - i)$. This is an excellent example of the fundamental importance of the coherence of the electron-pair wave in a superconductor.

We now find the value of the critical measuring current and how it depends on the strength of the magnetic field. Let α and β be the phase changes produced by currents flowing across the two weak-links (we must remember our sign convention: the clockwise direction is positive, so a positive phase difference is an increase in phase angle ϕ in a clockwise direction round the ring), and let $\Delta\phi(B)$ be the total phase change produced by the applied magnetic field around the top and bottom halves of the ring. If the ring is superconducting,

$$\alpha + \beta + \Delta\phi(B) = n \cdot 2\pi.$$

Equation (11.23) gives $\Delta\phi(B)$ in terms of the flux Φ_a enclosed by the ring; so

i.e. $$\alpha + \beta + 2\pi \frac{\Phi_a}{\Phi_0} = n \cdot 2\pi. \qquad (11.26)$$

This relation must always be satisfied if there is to be superconductivity everywhere round the ring. The magnitudes of α and β depend on the total current passing through the weak-links so, when the current I is fed through the ring the circulating current i adjusts itself so that the phase condition (11.26) is still satisfied.

In the absence of the measuring current I the phase differences across the two weak links are equal because the same current i flows through each. From (11.26) we see that in this case $\alpha = \beta = \pi[n - (\Phi_a/\Phi_0)]$. But when we pass the current I, α and β are no longer equal, because the current through X is now $i - \frac{1}{2}I$ and that through W is now $i + \frac{1}{2}I$. Since $\alpha + \beta$ must remain constant, the decrease in α must equal the increase in β. Let us write that when the measuring current I is passed from Y to Z

$$\alpha = \pi[n - (\Phi_a/\Phi_0)] - \delta,$$
$$\beta = \pi[n - (\Phi_a/\Phi_0)] + \delta \qquad (11.27)$$

where δ depends on the current I. If the weak-links are Josephson tunnelling junctions the phase difference across them due to current through them is given by (11.14.i), so for the weak-links X and W we have

COHERENCE OF THE ELECTRON-PAIR WAVE; QUANTUM INTERFERENCE 177

$$i - \tfrac{1}{2}I = i_c \sin\left[\pi\left(n - \frac{\Phi_a}{\Phi_0}\right) - \delta\right],$$
(11.28)
$$i + \tfrac{1}{2}I = i_c \sin\left[\pi\left(n - \frac{\Phi_a}{\Phi_0}\right) + \delta\right].$$

We are interested in how the SQUID behaves as we increase I, so we eliminate i by subtraction,

$$I = i_c\left\{\sin\left[\pi\left(n - \frac{\Phi_a}{\Phi_0}\right) + \delta\right] - \sin\left[\pi\left(n - \frac{\Phi_a}{\Phi_0}\right) - \delta\right]\right\}$$
$$= 2i_c \cos\left[\pi\left(n - \frac{\Phi_a}{\Phi_0}\right)\right]\sin\delta. \qquad (11.29)$$

The right-hand side of this equation could have either positive or negative values, depending on whether the cos and sin terms have the same or opposite signs. But we are considering what happens when the measuring current I is fed through from Y to Z in Fig. 11.8, and we have taken I in this direction to be positive. If I turns out to be negative, this is equivalent to turning the figure upside down. This inversion does not, however, represent a physically different situation, so we may write (11.29) as

$$I = 2i_c\left|\cos\left[\pi\left(n - \frac{\Phi_a}{\Phi_0}\right)\right]\sin\delta\right|,$$

i.e.
$$I = 2i_c\left|\cos\pi\frac{\Phi_a}{\Phi_0}\cdot\sin\delta\right|. \qquad (11.30)$$

However, $\sin\delta$ cannot have a magnitude greater than unity, so (11.30) can only be satisfied and all of the ring remain superconducting if

$$I \leqslant 2i_c\left|\cos\pi\frac{\Phi_a}{\Phi_0}\right|.$$

Therefore the critical measuring current is

$$I_c = 2i_c\left|\cos\pi\frac{\Phi_a}{\Phi_0}\right|. \qquad (11.31)$$

It can be seen from this that, when a magnetic field is applied, the critical measuring current of the SQUID depends in a periodic manner on the strength of the field, being a maximum when the field is such that the flux Φ_a through the ring is an integral number of fluxons (Fig. 11.7(c)).

A graph such as Fig. 11.7(c), which shows the periodic variation of critical current with applied magnetic field strength, is referred to as the "interference pattern" of the superconducting quantum interferometer.

In order to obtain a pattern such as Fig. 11.7(c) the critical current of the links and the area of the hole must be small, so that $Li_c \ll \Phi_0$ and the induced current cannot appreciably screen the hole from the applied field. Often, however, the critical currents of the links and the size of the hole may be large enough so that $Li_c > \tfrac{1}{2}\Phi_0$, and there is appreciable screening of the hole. It can be shown that in this case the depth of modulation (Fig. 11.9) of the critical current is reduced to $\Delta I_c = \Phi_0/L$.

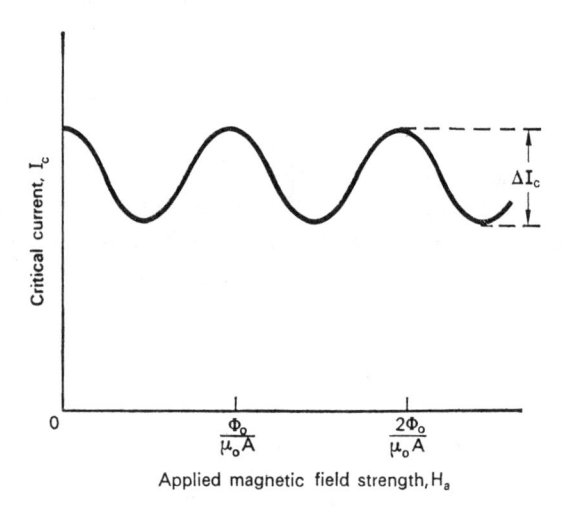

FIG. 11.9. Interference pattern of superconducting quantum inteferometer ($Li_c > \Phi_0$).

11.4.1. "Diffraction" effects

We have seen that the critical current of a SQUID is modulated by an applied magnetic field. It also happens, however, that the critical current of a *single weak-link* is modulated by a magnetic field. Although, as we showed in the last section, a magnetic field produces negligible *phase difference* across the weak-links in a SQUID, it nevertheless has a considerable effect on their *critical current*. It can be shown that the critical current i_c of a Josephson junction in a magnetic field parallel to the plane of the junction is itself a periodic function of the strength of the magnetic field, being a minimum when the magnetic flux passing through the junction equals an integral number of fluxons.

Equation (11.31) gives the critical current of a SQUID, i.e. two weak-links in parallel. This equation has exactly the same form as the Fraunhofer formula for the optical interference pattern from two parallel slits (Young's fringes) if we take I_c as equivalent to the resultant amplitude A of the optical oscillation, i_c as the amplitude a of each of the two interfering beams and $\pi\Phi_a/\Phi_0$ as half the phase difference between the two interfering beams. The amplitude of Young's fringes is, however, modulated by the diffraction pattern of the individual slits. The Fraunhofer formula for optical diffraction from a single slit is

$$a = a_0 \frac{\sin \beta}{\beta}$$

where β is one-half of the phase difference of rays from the two opposite edges of the slit. By analogy, therefore, we might expect the critical current i_c of a single weak-link to depend on the magnetic field in the following way:

$$i_c = i_c(0) \left| \frac{\sin(\pi\Phi_J/\Phi_0)}{\pi\Phi_J/\Phi_0} \right| \qquad (11.32)$$

where Φ_J is the flux of the applied magnetic field passing through the area of the link. In fact, a direct calculation, similar to that in the previous section, of the phase of the electron-pair wave in the neighbourhood of a weak-link leads to exactly this result.†

Just as optical interference patterns which are formed by a number of slits are modulated in amplitude by the diffraction which occurs at each individual slit, the interference pattern from a superconducting interferometer is modulated by the "diffraction" occurring at the weak-links. We neglected this diffraction effect in drawing Figs. 11.7(c) and 11.9. An actual interference pattern obtained from a quantum interferometer is shown in Fig. 11.10. The modulation of the interference pattern by the diffraction pattern of the weak-links is evident.

It is very difficult to make a superconducting quantum interferometer in which both weak-links have the same cross-sectional area. Consequently the interference pattern is usually modulated by two periodicities and a complicated multiply-periodic pattern results.

We have seen that there is a close analogy between quantum interference effects in superconductors and optical interference. Nevertheless, the reader should be warned against drawing false conclusions

† The detailed calculation may be found in Jaklevic, Lambe, Mercereau and Silver, *Phys. Rev.* **140A**, 1628 (1965).

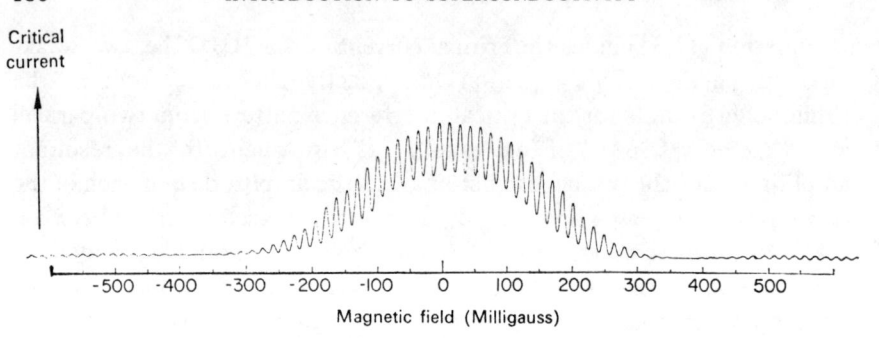

FIG. 11.10. Trace of interference pattern from a quantum interferometer, showing modulation of interference by diffraction. (Reproduced by kind permission of R. C. Jaklevic, J. Lambe, J. E. Mercereau, and A. H. Silver, Scientific Laboratory, Ford Motor Company.)

from this analogy about the nature of superconducting quantum interference. It must be remembered that the physical situations are quite different. The essential feature of optical interference is that interference between two coherent light waves produces at any point a resultant oscillation whose amplitude depends on the relative phase and amplitude of the two interfering waves at that point. This leads to a pattern in space whose intensity varies from place to place. In any uniform piece of superconducting metal, however, the intensity (i.e. square of amplitude) of the coherent electron-pair wave is everywhere the same. This is because the intensity is proportional to the density of superelectrons which has negligible variation from place to place. Equation (11.32), though of the same mathematical form as the equation of optical diffraction, does not contain any position variables and does not describe any variation of intensity in space. The SQUID described in § 11.4 is not to be regarded as an analogue of Young's slits in which the weak-links correspond to two slits whose diffracted waves interfere at Z.

Superconducting quantum interference can be put to practical use in a number of ways. Most of the applications make use of the fact that a superconducting interferometer can be used to detect very small changes of magnetic field strength. We have seen that the critical current of the device undergoes a complete cycle for a change of magnetic flux of one fluxon through the enclosed hole, and because the fluxon is only about 2×10^{-15} weber (2×10^{-7} gauss cm^2) a very small change in the applied magnetic field strength can be observed as a change in critical current. As well as being used directly as magnetometers, SQUIDS can be used as very sensitive galvanometers, because any electric current must generate a magnetic field and this field can be detected by a SQUID.

PART II

TYPE-II SUPERCONDUCTIVITY

CHAPTER 12

THE MIXED STATE

FOR MANY years it was supposed that the behaviour which we have described in Part I of this book was characteristic of all superconductors. It had, indeed, been noticed that certain superconductors, especially alloys and impure metals, did not behave quite in the expected way, but this anomalous behaviour was usually ascribed to impurity effects, not considered to be of great scientific interest, and consequently little effort was made to understand it. However, in 1957 Abrikosov published a theoretical paper pointing out that there might be another class of superconductors with somewhat different properties, and it is now realized that the apparently anomalous properties of certain superconductors are not merely trivial impurity effects but are the inherent features of this other class of superconductor now known as "type-II".

One of the characteristic features of the type-I superconductors we considered in the first part of this book is the Meissner effect, the cancellation within the metal of the flux due to an applied magnetic field. We mentioned in § 6.7 that the occurrence of this perfect diamagnetism implies the existence of a surface energy at the boundary between any normal and superconducting regions in the metal. This surface energy plays a very important role in determining the behaviour of a superconductor; for example, as we shall now see, it determines whether the material is a type-I or a type-II superconductor.

Consider a superconducting body in an applied magnetic field of strength less than the critical value H_c, and suppose that within the material a normal region were to appear with boundaries lying parallel to the direction of the applied magnetic field. The appearance of such a normal region would change the free energy of the superconductor, and we may consider two contributions to this free energy change: a contribution arising from the bulk of the normal region and a contribution due to its surface. As we saw in Chapter 4, in an applied magnetic field of strength H_a, the free energy per unit volume of the normal state is greater than that of the superconducting, perfectly diamagnetic, state by an amount $\frac{1}{2}\mu_0(H_c^2 - H_a^2)$. Furthermore, as shown in Chapter 6, there is a surface energy associated with the boundary between a normal and a superconducting region. For the type-I superconductors we considered in the first part of this book this surface energy is positive. Hence, if a normal region were to form in the superconducting material, there would be an increase in free energy due both to the bulk and to the surface of the normal region. For this reason, the appearance of normal regions is energetically unfavourable, and a type-I superconductor remains superconducting throughout when a magnetic field of strength less than H_c is applied.

Suppose, however, that in certain metals the surface energy between normal and superconducting regions were *negative* instead of positive (i.e. energy is released when the interface is formed). In this case the appearance of a normal region would reduce the free energy, if the increase in energy due to the bulk of the region were outweighed by the decrease due to its surface. A material assumes that condition which has the lowest total free energy, so in the case of a sufficiently negative surface energy we would expect that, in order to produce the minimum free energy, a large number of normal regions would form in the superconducting material when a magnetic field is applied. The material would split into some fine-scale mixture of superconducting and normal regions whose boundaries lie parallel to the applied field, the arrangement being such as to give the maximum boundary area relative to volume of normal material. We shall call this the *mixed state*. In the next section it will be shown that the conditions in some superconductors are such that the surface energy is indeed negative. These metals are therefore able to go into the mixed state, and these are the type-II superconductors.

It is important to distinguish clearly between the *mixed* state which occurs in type-II superconductors and the *intermediate* state which occurs in type-I superconductors. The intermediate state occurs in those type-I superconducting bodies which have a non-zero demagnetizing factor, and its appearance depends on the shape of the body. The mixed state, however, is an intrinsic feature of type-II superconducting material and appears even if the body has zero demagnetizing factor (e.g. a long rod in a parallel field). In addition, the structure of the intermediate state is relatively coarse and the gross features can be made visible to the naked eye [Figs. 6.6 and 6.7 (pp. 73 and 74)]. The structure of the mixed state is, as we shall see, on a much finer scale with a periodicity less than 10^{-5} cm (see *frontispiece*).

12.1. Negative Surface Energy

In § 6.9 we showed that, as a result of the existence of the penetration depth and coherence length, there is a surface energy associated with the boundary between a normal and superconducting region. It was shown that, if the coherence range is *longer* than the penetration depth, as it is in most pure metallic elements, the total free energy is increased close to the boundary [Fig. 6.9 (p. 78)], that is to say there is a *positive* surface energy.

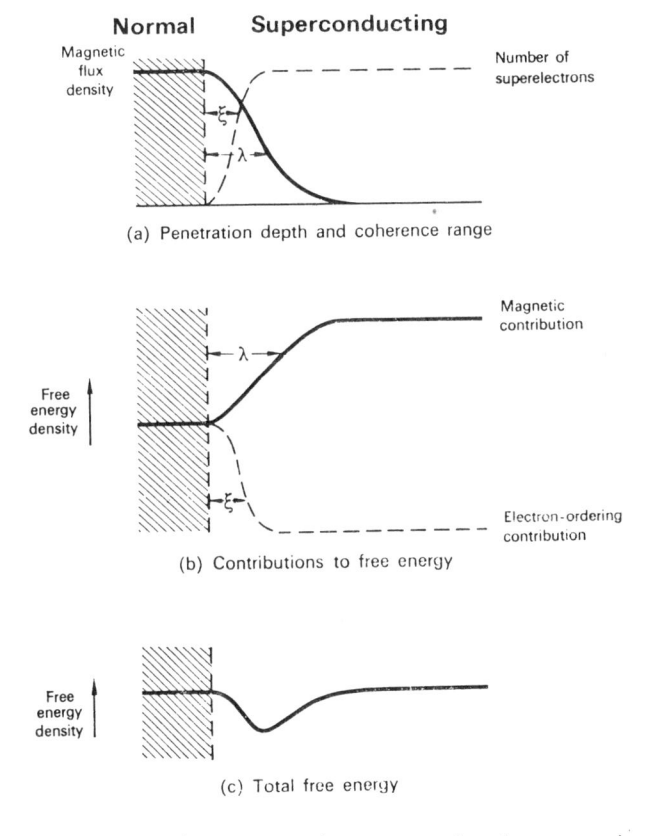

FIG. 12.1. Negative surface energy; coherence range less than penetration depth. (Compare this with Fig. 6.9.)

The relative values of the coherence length ξ and the penetration depth λ vary for different materials. In many alloys and a few pure metals the coherence range is greatly reduced, as was pointed out in Chapter 6. A similar argument to that used in § 6.9 shows that, if the coherence length is *shorter* than the penetration depth, the surface energy is *negative*, as Fig. 12.1 illustrates, and therefore such a superconductor will be type-II. In most pure metals the coherence length has a value ξ_0 of about 10^{-4} cm. This is considerably greater than the penetration depth, which is about 5×10^{-6} cm, so in such metals the surface energy is positive and they are type-I. A reduction in the electron mean free path, however, reduces the coherence length and increases the penetration depth (§ 6.9, § 2.4.1). Impurities in a metal reduce the electron mean free path, and in an impure metal or alloy the coherence range can easily be less than the penetration depth. Alloys or sufficiently impure metals are, therefore, usually type-II superconductors.

12.2. The Mixed State

We have seen that it may be energetically favourable for superconductors with a negative surface energy between normal and superconducting regions to go into a mixed state when a magnetic field is applied. The configuration of the normal regions threading the superconducting material should be such that the ratio of surface to volume of normal material is a maximum. It turns out that a favourable configuration is one in which the superconductor is threaded by cylinders of normal material lying parallel to the applied magnetic field (Fig. 12.2). We shall refer to these cylinders as *normal cores*. These cores arrange themselves in a regular pattern, in fact a triangular close-packed lattice (Fig. 12.2 and *frontispiece*).

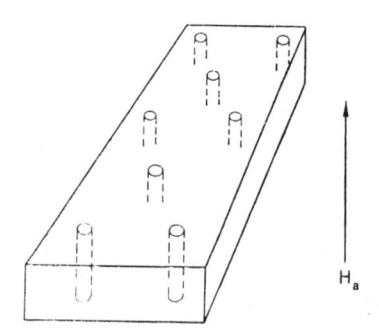

FIG. 12.2. The mixed state.

We might expect the normal cores to have a very small radius because the smaller the radius of a cylinder the larger the ratio of its surface area to its volume. The picture of the mixed state which emerges from these considerations is as follows. The bulk of the material is diamagnetic, the flux due to the applied field being opposed by a diamagnetic surface current which circulates around the perimeter of the specimen. This diamagnetic material is threaded by normal cores lying parallel to the applied magnetic field, and within each core is magnetic flux having the same direction as that of the applied magnetic field. The flux within each core is generated by a vortex of persistent current that circulates around the core with a sense of rotation opposite to that of the diamagnetic surface current. (We saw in § 2.3.1 that any normal region containing magnetic flux and enclosed by superconducting material must be encircled by such a current.) The pattern of currents and the resulting flux are illustrated in Fig. 12.3.

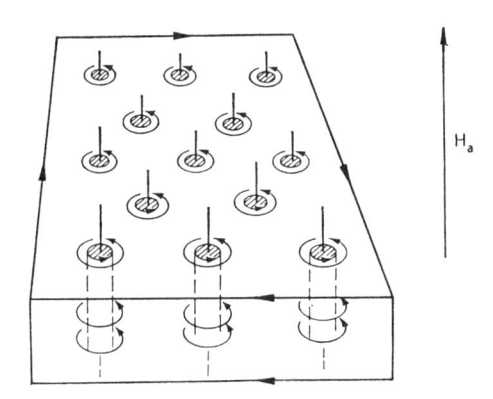

FIG. 12.3. The mixed state, showing normal cores and encircling supercurrent vortices. The vertical lines represent the flux threading the cores. The surface current maintains the bulk diamagnetism.

The vortex current encircling a normal core interacts with the magnetic field produced by the vortex current encircling any other core and, as a result, any two cores repel each other. This is somewhat similar to the repulsion between two parallel solenoids or bar magnets. Because of this mutual interaction the cores threading a superconductor in the mixed state do not lie at random but arrange themselves into a regular periodic hexagonal array† as shown in Fig. 12.3. This array is usually known as the *fluxon lattice*. The existence of the normal cores and their arrangement in a periodic lattice has been revealed by two experimental techniques. The decoration technique of Essmann and

† Occasionally a square lattice may be formed, but this is very uncommon and only occurs under special circumstances.

Träuble reveals the pattern of the normal cores by allowing very small (500 Å) ferromagnetic particles to settle on the surface of a type-II superconductor in the mixed state. The particles locate themselves where the magnetic flux is strongest, i.e. where the normal cores intersect the surface. The resulting pattern can then be examined by electron microscopy. The frontispiece shows the pattern of normal cores in the mixed state revealed by this method. An alternative method, developed by Cribier, Jacrot, Rao and Farnoux, makes use of neutron diffraction. Neutrons, because of their magnetic moment, interact with magnetic fields. The regular arrangement of current vortices in the mixed state produces a periodic magnetic field which acts as diffraction grating, and scatters a beam of neutrons shone through the specimen into preferential directions given by the Bragg law. Observation of the directions into which the beam of neutrons is scattered shows that the cores are arranged in a hexagonal periodic array.

12.2.1. Details of the mixed state

The picture of the mixed state we have just given, with thin cylindrical normal cores threading the superconducting material, is a good enough approximation for many purposes, but it does not accurately describe the details of the structure. For one thing, the cores are not sharply defined. We saw in § 6.9 that there cannot be a sharp boundary between a superconducting and a normal region; the transition is spread out over a distance which is roughly equal to the coherence range ξ. Furthermore, the magnetic flux associated with each core spreads into the surrounding material over a distance about equal to the penetration depth λ.

A detailed analysis of the free energy of the mixed state, which we shall not attempt here, shows that the normal cores should have an exceedingly small radius. However, as the size of a normal cylinder is reduced so that it approaches ξ it becomes progressively more difficult to define its radius exactly on account of the diffuseness of the boundary. Because it is not possible to define exactly the volume and surface area of such a small core, we cannot properly divide its free energy into distinct volume and surface contributions, and we must consider its free energy as a whole. A detailed consideration of free energy gives the following structure for the mixed state.

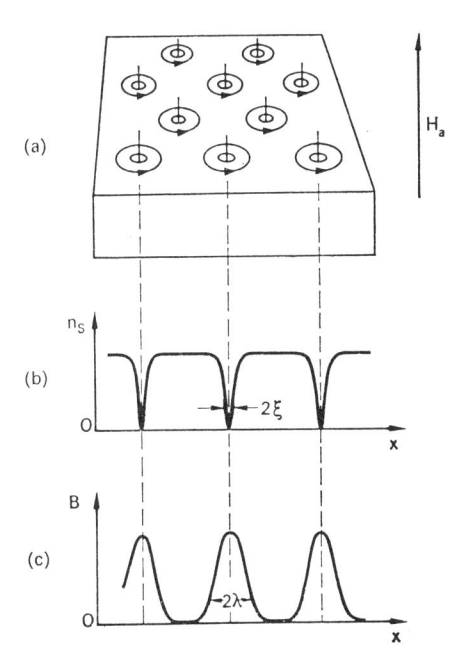

FIG. 12.4. Mixed state in applied magnetic field of strength just greater than H_{c1}. (a) Lattice of cores and associated vortices. (b) Variation with position of concentration of superelectrons. (c) Variation of flux density.

The properties of the material vary with position in a periodic manner. Towards the centre of each vortex the concentration n_s of superelectrons falls to zero, so along the centre of each vortex is a very thin core (strictly a line) of normal material (Fig. 12.4b). The dips in the superelectron concentration are about two coherence-lengths wide. The flux density due to the applied magnetic field is not cancelled in the normal cores and falls to a small value over a distance about λ away from the cores (Fig. 12.4c). The total flux generated at each core by the encircling current vortex is just one fluxon (see § 11.2).

We shall now confirm that, when a magnetic field is applied to a type-II superconductor, the appearance of cores of the form we have just described does result in a lowering of the free energy. At each core the number n_s of superelectrons decreases and energy must be provided to split up the pairs. As an approximation we may think of each core as equivalent to a cylinder of normal material with radius ξ. The appearance of a normal core will therefore result in a local increase in free energy of $\pi\xi^2 . \frac{1}{2}\mu_0 H_c^2$ per unit length of core due to the decrease in electron order. However, over a radius of about λ the material is not diamagnetic so there is a local decrease in magnetic energy approximately equal to $\pi\lambda^2 \cdot \tfrac{1}{2}\mu_0 H_a^2$ per unit length, where H_a is the strength of the applied field. If there is to be a net reduction in free energy by the formation of such cores, we must have

$$\pi\xi^2 \cdot \tfrac{1}{2}\mu_0 H_c^2 < \pi\lambda^2 \cdot \tfrac{1}{2}\mu_0 H_a^2. \tag{12.1}$$

According to this relation, if the mixed state is to appear in applied fields less than H_c (a necessary condition, otherwise an applied field would drive the whole superconductor into the normal state before the mixed state could establish itself), we must have $\xi < \lambda$. This is the same condition as that we derived for negative surface energy. So, as predicted by the simple arguments on page 184, the mixed state is produced by the application of a magnetic field to superconductors which would have negative surface energy between superconducting and normal regions.

12.3. Ginzburg–Landau Constant of Metals and Alloys

Let us write the ratio of the penetration depth λ to the coherence length ξ as a parameter $\varkappa$:

$$\varkappa = \lambda/\xi.$$

$\varkappa$ varies for different superconductors and is known as the Ginzburg–Landau constant of the material.† It is an important parameter because its value determines several properties of the superconductor; for example, according to the considerations of the previous sections, a superconductor is type-I or type-II depending on whether its value of $\varkappa$ is less or greater than unity.

A more detailed treatment than we have given shows that the sign of the surface energy and the possibility of the formation of a mixed state depends, strictly, not on whether the $\varkappa$ of the material is less or greater than unity, but on whether $\varkappa$ is less or greater than $1/\sqrt{2}$:

$\varkappa < 0{\cdot}71$ surface energy positive (type-I),

$\varkappa > 0{\cdot}71$ surface energy negative (type-II).

---

† The constant $\varkappa$ appears in the theoretical treatment of superconductivity by Ginzburg and Landau, which is an extension of the London treatment and explicitly includes the surface energy between normal and superconducting regions. In the Ginzburg–Landau treatment $\varkappa$ is defined by $\varkappa = (\sqrt{2})2\pi\lambda^2\mu_0 H_c/\Phi_0$ where Φ_0 is the quantum of magnetic flux (see § 11.2). If the electron mean free path is very short, λ increases and so $\varkappa$ is large in alloys. For our purposes, however, we can consider $\varkappa$ to be the ratio of the penetration depth to the coherence range.

This correct critical value of $\varkappa$ is, however, not very different from the value of unity we obtained by simple considerations.

It was pointed out in § 12.1 that in alloys and impure metals the coherence range is shorter than in pure metals; consequently $\varkappa$ can have a large value and these superconductors are usually type-II. It is, however, possible for even pure metals to be type-II superconductors. It can be shown that superconductors with high transition temperatures can be expected to have relatively short coherence ranges and, in fact, three superconducting metals (niobium, vanadium, and technetium) have $\varkappa$ greater than 0·71 even in the absence of impurities. These are called *intrinsic* type-II superconductors. Pure niobium, vanadium, and technetium have $\varkappa$ values of 0·78, 0·82 and 0·92, respectively.† However, pure metals are usually type-I and alloys are usually type-II.

The Ginzburg–Landau constant $\varkappa$ of a superconductor which contains impurities is related to its resistivity in the normal state because the scattering of electrons by the impurities shortens the coherence range ξ and also increases the normal resistivity ρ. For a given metal, therefore, $\varkappa$ increases with the normal state resistivity.

12.4. Lower and Upper Critical Fields

12.4.1. Lower critical field, H_{c1}

We have seen that when a magnetic field is applied to a type-II superconductor it may be energetically favourable for it to go into the mixed state whose configuration has been described in previous sections. However, a certain minimum strength of applied field is required to drive a type-II superconductor into the mixed state. This can be seen by examination of (12.1), which gives the condition for the free energy to be lowered by the appearance of the mixed state. For a given value of ξ relative to λ (remembering that, in a type-II superconductor, $\xi < \lambda$), we see that H_a must be greater than a certain fraction of H_c. Therefore a certain minimum strength of applied field is required to drive a type-II superconductor into the mixed state, and this is known as the *lower critical field*, H_{c1}. We can get an approximate value for H_{c1} from eqn.

---

† In fact, for intrinsic type-II superconductors and dilute alloys $\varkappa$ varies slightly with temperature, increasing as the temperature falls. In vanadium, for example, $\varkappa$ at the transition temperature (5·4°K) is 0·82 but rises to 1·5 at 0°K. Similarly the alloy Pb.$_{99}$Tl.$_{01}$ has a $\varkappa$-value of 0·58 at its transition temperature, 7·2°K, and so is type-I, but on cooling to 4·3°K the $\varkappa$-value rises to 0·71, so below this temperature the alloy is type-II.

(12.1) (approximate because eqn. (12.1) was based on a simplified model of a core). From this equation we see that the mixed state will be energetically favoured if the strength of the applied field exceeds $H_c \xi / \lambda$, i.e.

$$H_{c1} \simeq H_c / \varkappa.$$

Clearly the value of H_{c1} relative to H_c decreases as the value of $\varkappa$ increases.

12.4.2. Upper critical field, H_{c2}

In the previous section we have shown that if a gradually increasing magnetic field is applied to a type-II superconductor it goes into the mixed state at a "lower critical field" H_{c1} which is less than H_c. Now, in a type-I superconductor, H_c is the field strength at which the magnetic free energy of the superconductor has been raised to such an extent that it becomes energetically favourable for it to go into the normal state. A type-II superconductor in the mixed state has, however, in an applied field, a lower free energy than if it were type-I and perfectly diamagnetic. Consequently we may expect that a magnetic field stronger than H_c must be applied to drive a type-II superconductor normal. (This is similar to the argument used in § 8.1 to show that the critical field of a thin superconductor is greater than the critical field of the bulk material.) Furthermore, we may note that an argument similar to that on p. 189 shows that in fields *above* H_c the mixed state can have a lower free energy than the completely normal state. The high magnetic field strength up to which the mixed state can persist is called the *upper critical field*, H_{c2}.

At the lower critical field strength H_{c1} a type-II superconductor goes from the completely superconducting state into the mixed state and a lattice of parallel cores is formed. As the strength of the applied magnetic field is increased above H_{c1} the cores pack closer together and, because each core is associated with a fixed amount of flux, the average flux density B in the superconductor increases. At a sufficiently high value of applied magnetic field the cores merge together and the mean flux density in the material due to the cores and the diamagnetic surface current approaches the flux density $\mu_0 H_a$ of the applied magnetic field (Fig. 12.5). At the upper critical field H_{c2} the flux density becomes equal to $\mu_0 H_a$ and the material goes into the normal state.

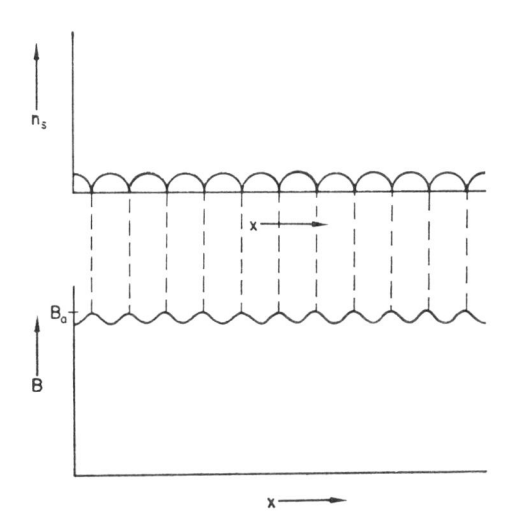

FIG. 12.5. Mixed state at applied magnetic field strength just below H_{c2}.

We have now seen that, whereas type-I superconductors can exist in one of two states, superconducting or normal, type-II superconductors can be in one of *three* states, superconducting, mixed or normal. The phase diagrams of the two types are compared in Fig. 12.6. In a type-II superconductor, the larger the value of $\varkappa$ the smaller will be H_{c1} but the larger will be H_{c2} relative to the critical field H_c.

12.4.3. Thermodynamic critical field, H_c

In Chapter 4 we saw that for a type-I superconductor the critical field has a value given by

$$H_c = \left[\frac{2}{\mu_0}(g_n - g_s)\right]^{\frac{1}{2}}, \qquad (12.2)$$

where $(g_n - g_s)$ is the difference in free energy densities of the normal and superconducting states in the absence of an applied magnetic field. We may *define* the critical field for *all* types of superconductor by means of (12.2), which can apply equally to type-I and type-II, since for each superconductor there must be, in the absence of an applied magnetic field, a characteristic energy difference $(g_n - g_s)$ between the completely superconducting and completely normal states. H_c is a measure of this energy difference and, to distinguish it from the upper and lower critical fields, it may be called the *thermodynamic critical field*. Only in a type-I

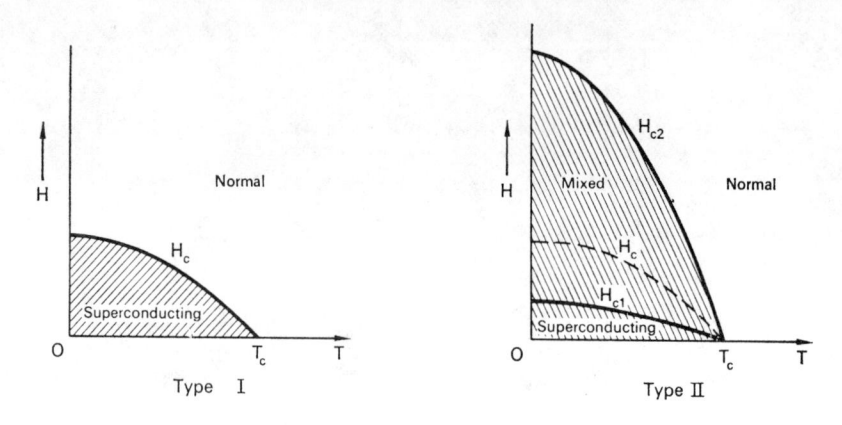

FIG. 12.6. Phase diagrams of type-I and type-II superconductors.

superconductor does the material become normal in a field strength equal to H_c.

When, in describing a type-II superconductor, we say that the upper or lower critical fields have certain values relative to the thermodynamic critical field H_c, we may loosely think of H_c as being about the critical field that would be characteristic of an equivalent type-I superconductor, i.e. one with the same transition temperature.† As will be seen in § 12.5.1, the value of H_c for a type-II superconductor can nevertheless be determined indirectly from its experimentally measured magnetization curve.

12.4.4. Value of the upper critical field

It can be shown that for a type-II superconductor the upper critical field has a value

$$H_{c2} = (\sqrt{2})\varkappa H_c, \qquad (12.3)$$

so materials with a high value of $\varkappa$ remain in the mixed state and do not go normal until strong magnetic fields are applied.

The ability of type-II superconductors with high values of $\varkappa$ to resist being driven normal until strong magnetic fields are applied is of considerable technical importance, especially in the construction of super-

---

† We saw in Chapter 9 that the BCS theory of superconductivity predicts a law of corresponding states for different superconductors, from which it follows that superconductors with the same transition temperature should have the same value of H_c at any temperature. This law of corresponding states is fairly well obeyed in practice.

conducting solenoids to generate strong magnetic fields. Whereas at 4·2°K type-I superconductors have critical fields of only a few times 10^4 A m^{-1} (i.e. a few hundred gauss), type-II superconductors can have upper critical field strengths exceeding a million A m^{-1}. Some typical examples are shown in Table 12.1.

TABLE 12.1. UPPER CRITICAL FIELD H_{c2} AT 4·2°K, $\varkappa$-VALUE AND TRANSITION TEMPERATURE OF SOME TYPE-II ALLOYS COMPARED TO LEAD (TYPE-I)

		H_{c2} (4·2°K)		$\varkappa$	T_c (°K)
		A m^{-1}	Gauss		
Type-II	Mo$_3$–Re	$6·7 \times 10^5$	8,400	4	10
	Ti$_2$–Nb	$\sim 8 \times 10^6$	$\sim 100,000$	20	9
	Nb$_3$Sn	$\sim 1·6 \times 10^7$	$\sim 200,000$	34	18
		H_c (4·2°K)			
Type-I	Pb	$4·4 \times 10^4$	550	0·4	7·2

12.4.5. Paramagnetic limit

The question may be asked, if $\varkappa$ is made indefinitely large, is there any limit to the strength of magnetic field required to drive a type-II superconductor normal? To answer this question, consider a material with a high transition temperature and a large value of $\varkappa$. At temperatures well below the transition temperature the thermodynamic critical field will be fairly high and so at these temperatures, according to (12.3), we should have a very large value of H_{c2}. For example, at 1·2°K an alloy of 60 atomic per cent titanium and 40 atomic per cent niobium is predicted to have an upper critical field strength H_{c2} of about 20×10^6 A m^{-1}. Experiment shows, however, that for such materials with a very high predicted value of H_{c2}, resistance in fact returns at a considerably weaker field. In the case of the titanium–niobium alloy, the normal state is restored by a magnetic field of strength 10×10^6 A m^{-1}, about half the predicted value of H_{c2}.

This reduced critical field has been ascribed to paramagnetism arising from the spins of the conduction electrons. A magnetic field applied to a normal metal tends to align parallel to itself the spins of the electrons near the Fermi level (Pauli paramagnetism). For moderate magnetic field strengths the degree of alignment and the resulting lowering of the free energy is small, and until now we have neglected it, considering a normal metal to be non-magnetic. But in very strong magnetic fields there may be a considerable reduction of magnetic free energy if the spins of the electrons align parallel to the applied field. Such an alignment is, however, incompatible with superconductivity, which requires that in each Cooper pair the spins of the two electrons shall be anti-parallel. Consequently, in a sufficiently strong magnetic field it may be energetically favourable for the metal to go into the normal paramagnetic state with electrons near to the Fermi level aligned parallel to the field, rather than to remain superconducting with electrons in anti-parallel pairs.

A calculation by Clogston suggests that as a result of the electron paramagnetism a superconductor must go into the normal state if the applied field strength exceeds a strength H_p, equal to about $1 \cdot 4 \times 10^6 \, T_c$ A m^{-1}. Consequently the mixed state of a type-II superconductor cannot persist in fields above this value, no matter how large the value of $\varkappa$.

The same limitation should apply to the increased critical field of a very thin specimen of type-I superconductor. Equation (8.8) implies that the critical field of a thin film could be made indefinitely high if the film were thin enough. In fact, however, the critical field will not increase above the value H_p.

It must be remembered that this effect of the normal electron paramagnetism is only important in superconductors with a very high upper critical field, say 5×10^6 A m^{-1} or more. The effect of electron paramagnetism is negligible in superconductors to which we only apply a relatively weak magnetic field.

To summarize: subject to the considerations which have been discussed in the previous paragraphs, the values of the lower and upper critical fields for a type-II superconductor with Ginzburg–Landau constant $\varkappa$ are given approximately by

$$H_{c1} \simeq \frac{1}{\varkappa} H_c,$$

$$H_{c2} \simeq \varkappa H_c.$$

12.5. Magnetization of Type-II Superconductors

We now examine the magnetic properties of type-II superconductors. At applied magnetic field strengths H_a below H_{c1}, a type-II superconductor behaves exactly like a type-I superconductor, exhibiting perfect diamagnetism and a magnetization equal to $-H_a$ (Fig. 12.7). When the

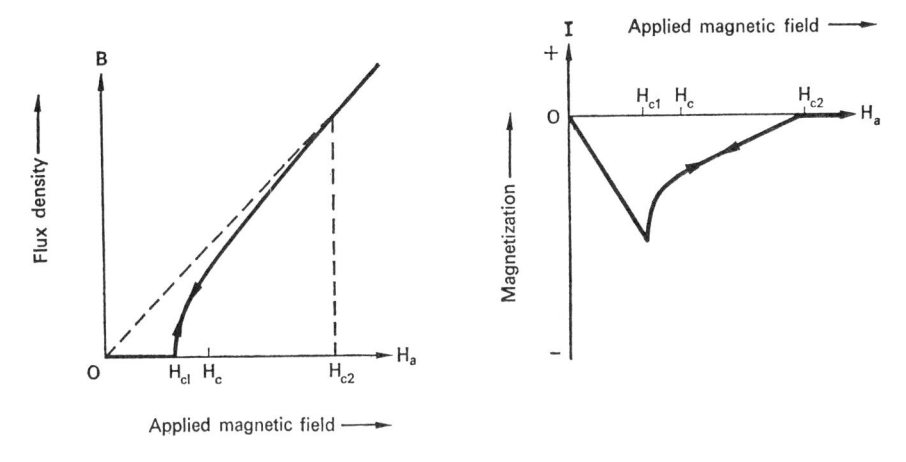

FIG. 12.7. Magnetization of a type-II superconductor.

applied field strength reaches H_{c1}, normal cores with their associated vortices form at the surface and pass into the material. The flux threading the vortices is in the same direction as that due to the applied magnetic field, so the flux in the material is no longer equal to zero and the magnitude of the magnetization suddenly decreases (Fig. 12.7). In fields between H_{c1} and H_{c2} the number of vortices which occupy the sample is governed by the fact that vortices repel each other. The number of normal cores per unit area for a given strength of applied magnetic field is such that there is equilibrium between the reduction in free energy of the material due to the presence of each non-diamagnetic core and the existence of the mutual repulsion between the vortices. As the strength of the applied field is increased, the normal cores pack closer together, so the average flux density in the material increases and the magnitude of the magnetization decreases smoothly with increasing H_a. Near to the upper critical value H_{c2} the flux density and magnetization change linearly with applied field strength. At H_{c2} there is a discontinuous change in the slope of the flux density and magnetization curves, and above H_{c2} the material is in the normal state with flux density equal to $\mu_0 H_a$ and zero magnetization.

We pointed out in § 4.1 that the total area enclosed by the magnetization curve is always equal to the difference between G_n and G_s, i.e. to $\frac{1}{2}\mu_0 H_c^2 V$, and this remains true for a type-II superconductor. In Fig. 12.8

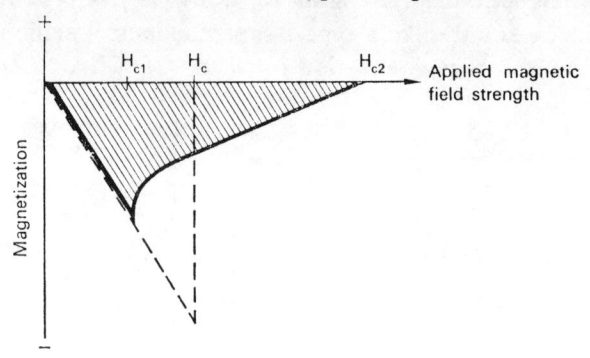

FIG. 12.8. Illustration of thermodynamic critical field H_c of type-II superconductor. The dotted right-angled triangle is drawn to have an area equal to the shaded area within the magnetization curve.

the relationship between H_{c2} and the thermodynamic critical field is illustrated. The ratio between H_{c2} and H_c is such that the area enclosed by the dashed curve equals the area enclosed by the magnetization curve of the type-II superconductor.

12.5.1. Determination of $\varkappa$

The value of $\varkappa$ of a type-II superconductor can be determined if a magnetization curve has been obtained. If the area under the magnetization curve is measured, the ratio of H_{c2} to H_c can be deduced by use of the construction shown in Fig. 12.8. Formula (12.3) then gives $\varkappa = (1/\sqrt{2})(H_{c2}/H_c)$. Furthermore, it has been shown that the slope of the magnetization curve where it cuts the applied field axis at H_{c2} is given by

$$\left[\frac{dI}{dH}\right]_{H_{c2}} = \frac{-1}{1 \cdot 16 \, (2\varkappa^2 - 1)},$$

so we can also determine $\varkappa$ from the slope of the measured magnetization curve. Note, however, that these procedures to determine $\varkappa$ are only valid if the magnetization is reversible (i.e. the same curve is traced in both increasing and decreasing fields). As we shall see in the next section, the magnetization is often not reversible, and the greater the degree of hysteresis the less accurate are the values of $\varkappa$ determined from the magnetization curve.

12.5.2. Irreversible magnetization

If a type-II superconductor is perfectly homogeneous in composition, its magnetization is reversible, i.e. the curves in Fig. 12.7 are the same whether the applied field H_a is increased from zero or decreased from some value greater than H_{c2}. Real samples, however, usually show some irreversibility in their magnetic characteristics (Fig. 12.9). Irreversibility

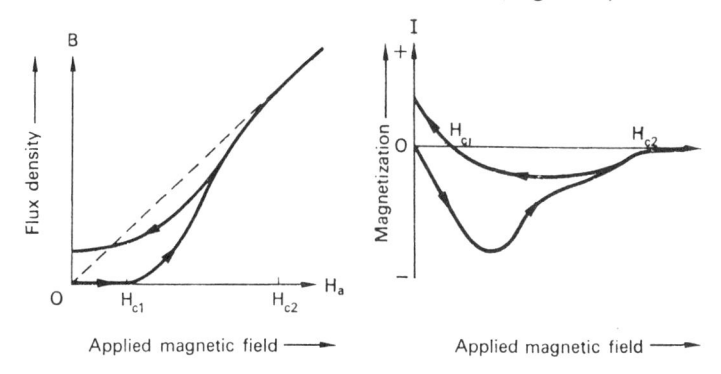

FIG. 12.9. Irreversible type-II magnetization curves.

is attributed to the fact that the normal cores which thread the superconductor in the mixed state, can be "pinned" to imperfections in the material and so are prevented from being able to move freely. Consequently, on increasing the applied field strength from zero there is no sudden entry of flux at H_{c1} because the cores formed at the surface are hindered from moving into the interior. Similarly, on reducing the applied field strength from a value greater than H_{c2}, there is a hysteresis, and flux may be left permanently trapped in the sample, because some of the normal cores are pinned and cannot escape. It appears that almost any kind of imperfection whose dimensions are as large or larger than the coherence length can pin normal cores. For example, both the long chains of lattice faults, called dislocations, and particles of chemical impurities, such as oxides, can give rise to magnetic irreversibility. But type-II materials which are very carefully prepared and purified so as to be free of such defects can show very little irreversibility and can exhibit magnetization curves almost as "ideal" as those in Fig. 12.7. In general, however, specimens contain a number of imperfections and their magnetization curves exhibit some irreversibility and permanently trapped flux. We shall see in the next chapter that the pinning of normal cores by imperfections plays a very important part in determining the critical currents of type-II superconductors.

12.6. Specific Heat of Type-II Superconductors

In § 5.2 we saw that if a type-I superconductor is heated there is, in general, a sudden change in the specific heat as the metal goes from the superconducting into the normal state; the magnitude of this change being given by (5.4). If the heating is done in a constant applied magnetic field H_a, the transition is of the first order, and latent heat is absorbed.

A type-II superconductor, however, has two critical fields, H_{c1} and H_{c2}. On heating a sample in a constant applied magnetic field H_a (e.g. from A to B in Fig. 12.10) we may expect to observe two changes in the specific heat, first at T_1 and then at T_2.

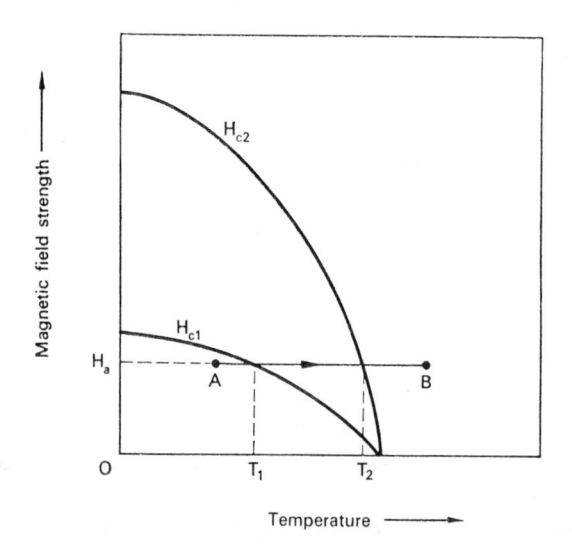

FIG. 12.10. Type-II superconductor heated in a magnetic field.

At temperature T_1 the metal passes from the superconducting to mixed state. It can be seen from Fig. 12.11 that this transition is associated with a large narrow peak in the specific heat curve. According to the Abrikosov model of a type-II superconductor, the magnetization curve at H_{c1} has an infinite slope (see Fig. 12.7), from which it can be shown that at T_1 the entropy is continuous but has an infinite temperature derivative in the mixed state. The lack of discontinuity in the entropy leads to a second-order transition and the infinite temperature derivative should produce a "λ-type" specific heat anomaly. The sharp peak observed at T_1 in Fig. 12.11 is consistent with such a

THE MIXED STATE 201

λ-type specific heat anomaly. At T_2 the specimen passes from the mixed to normal state. According to the description of the mixed state given in § 12.4.2 we would expect that at T_2 the nature of the mixed state approaches that of the normal state. We do not therefore expect any sudden increase in entropy as the metal is heated through T_2 but expect that the entropy of the mixed state will rise towards that of the normal state as the temperature is raised towards T_2. Hence this transition should also be of second order, and we expect at T_2 a sudden drop in the specific heat similar to that observed in type-I superconductors in the absence of a magnetic field. It can be seen in Fig. 12.11 that at T_2 such a specific heat drop is indeed observed.

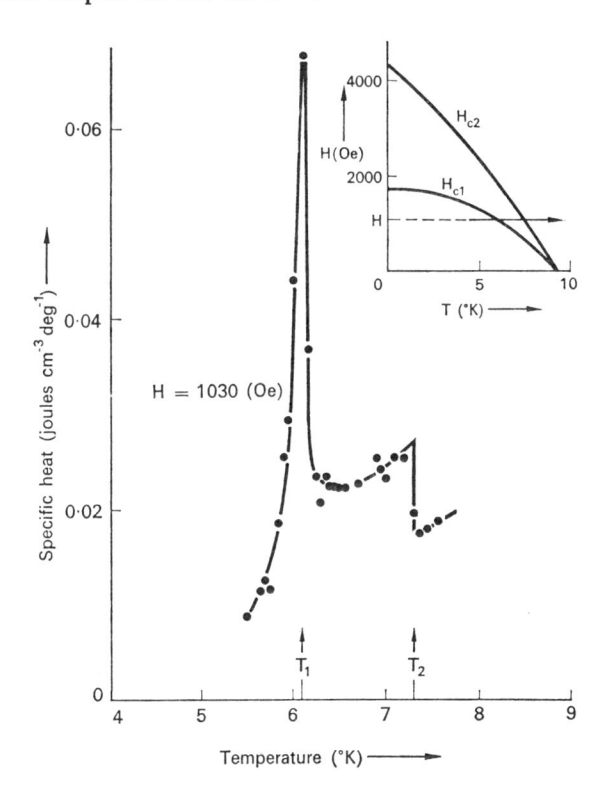

FIG. 12.11. Specific heat of type-II superconductor (niobium) measured in a constant applied magnetic field. (Based on McConville and Serin.)

CHAPTER 13

CRITICAL CURRENTS OF TYPE-II SUPERCONDUCTORS

THE CRITICAL currents of type-II superconductors are of considerable practical interest. We have mentioned previously that electromagnets capable of generating strong magnetic fields can be wound from wires of type-II superconductors, and clearly the more current that can be passed through the windings of such an electromagnet without resistance appearing the stronger will be the magnetic field that can be generated without heat being produced.

In Chapter 7 we saw that, provided the specimen is considerably larger than the penetration depth, the critical current of a type-I superconductor is successfully predicted by Silsbee's hypothesis, i.e. if the resistance is to remain zero, the total magnetic field strength at the surface, due to the current and applied magnetic field together, must not exceed H_c. The situation in type-II superconductors is, however, more complicated, because the state of the material changes at two field strengths, H_{c1} and H_{c2}, not at a single field strength H_c.

It should be pointed out that at present (1977) the behaviour of currents in type-II superconductors is by no means fully understood. Consequently we shall only discuss rather general aspects of the current-carrying capacity and we shall not try to present any detailed treatment, because present ideas are almost bound to be modified by future developments.

13.1. Critical Currents

In a magnetic field whose strength is less than H_{c1} a type-II superconductor is in the completely superconducting state and behaves like a type-I superconductor, whereas in field strengths greater than H_{c1} it goes into the mixed state. We shall see later that, contrary to what one might expect, the mixed state does not necessarily have zero resistance. We might guess, therefore, that, so long as the applied magnetic field is not by itself strong enough to drive the material into the mixed state, the critical current should be determined by a criterion like Silsbee's rule (§ 7.1) for a type-I superconductor, but with H_{c1} substituted for H_c; i.e. the metal will be resistanceless so long as the magnetic field generated by the transport current does not bring the total field at the surface to a value above H_{c1}.

Experiments show that for weak applied magnetic field strengths this modified Silsbee's rule is obeyed, but only in the case of extremely perfect samples, i.e. those with reversible magnetization curves. In the case of a field applied at right angles to a wire of pure type-II superconductor, the critical current falls linearly with increasing field strength, as would be expected (compare curve a Fig. 13.1 and Fig. 7.1b, p. 84). However, the critical current does not fall to zero at $\frac{1}{2}H_{c1}$, but there remains a small critical current; and even above H_{c1}, where the applied field is by itself strong enough to drive the metal into the mixed state, a type-II superconductor can still carry some resistanceless current. Curve a on Fig. 13.1 shows this small critical current extending up to about

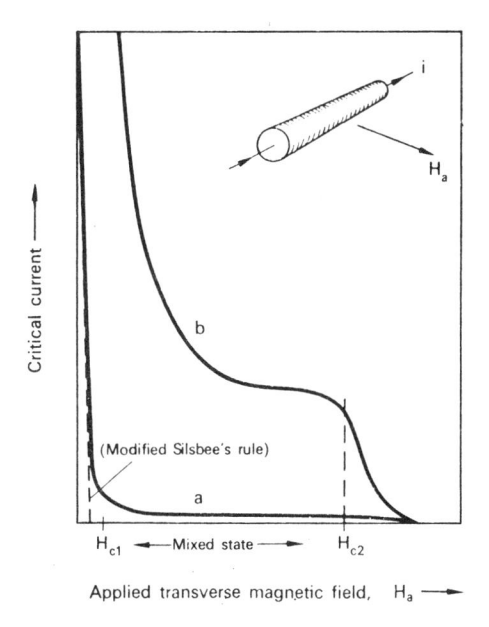

FIG. 13.1. Typical variation of critical currents of wires of (a) highly perfect, (b) imperfect type-II superconductors in transverse applied magnetic field.

H_{c2}. Most samples are not, however, extremely perfect, and for such imperfect samples the critical current is increased considerably both above and below H_{c1} (curve b, Fig. 13.1).

In this chapter we shall be chiefly concerned with the critical currents when the applied magnetic field is perpendicular to the current flow. There is particular interest in this configuration because it is the one which occurs in electromagnets. In a solenoid, for example, the field generated is perpendicular to the coil windings. The critical current curves of Fig. 13.1 refer to this situation. The shape of curve b, with a plateau extending up to about H_{c2}, is characteristic of type-II superconductors which are imperfect,† and is quite unlike that which would be predicted by any form of Silsbee's rule. It is found that when a superconductor is in the mixed state its critical current is almost completely controlled by the perfection of the material; the more imperfect the material the greater is the critical current, i.e. the higher and more pronounced is the plateau (Fig. 13.1). A highly imperfect wire may be able to carry about 10^3 A mm^{-2} of its cross-section. Conversely, a rather perfect specimen has a very small critical current, perhaps a few tens of microamps per mm^2, when it is in the mixed state. It is extremely important to understand that the critical current of a type-II superconductor in the mixed state is entirely determined by imperfections and impurities and not by any form of Silsbee's rule. This dependence of the critical current on the perfection of the material is of considerable technical importance because superconducting electromagnets require resistanceless wire of high current-carrying capacity. If Silsbee's rule applied, with H_{c2} as the appropriate magnetic field, the critical currents would be orders of magnitude greater than those which are actually found.

It sometimes happens that at high applied magnetic field strengths the critical current *increases* with increasing field strength, rising to a maximum near H_{c2}. This is known as the "peak effect". However, the reason for the occasional appearance of this effect is not yet understood, and we shall not consider it further in this book.

13.2. Flow Resistance

Before attempting an explanation of what determines how much resistanceless current can flow through a type-II superconductor when it is in the mixed state, we must draw attention to an important experimental observation. Suppose we take a length of wire of type-II

---

† When speaking of the perfection of a material we mean the lack of both "chemical" impurities (i.e. foreign atoms) and "physical" impurities (i.e. faults in the periodic arrangement of the atoms in the crystal lattice).

superconductor and apply perpendicular to it a magnetic field H_a of sufficient strength to drive the material into the mixed state. We now pass a current along the wire and observe how the voltage V developed between the ends varies as the magnitude i of the current is altered (Fig. 13.2). So long as the current is less than the critical value i_c no voltage is

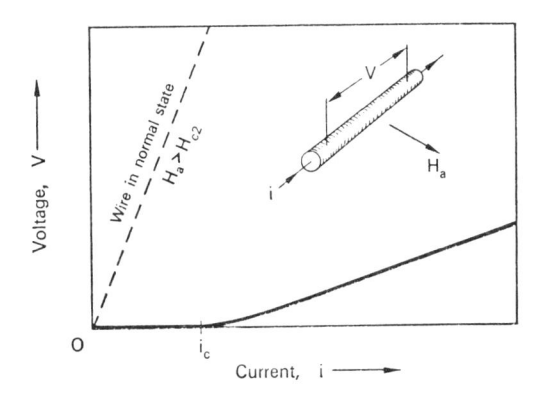

FIG. 13.2. Voltage–current characteristic of type-II superconductor in transverse magnetic field ($H_{c1} < H_a < H_{c2}$).

observed along the wire, but when the current is increased above i_c a voltage appears which, at currents somewhat greater than i_c, approaches a linear increase with increasing current. It should be noted that the voltage developed is considerably less than that which would be observed if the wire were in the normal state. Figure 13.3 shows the

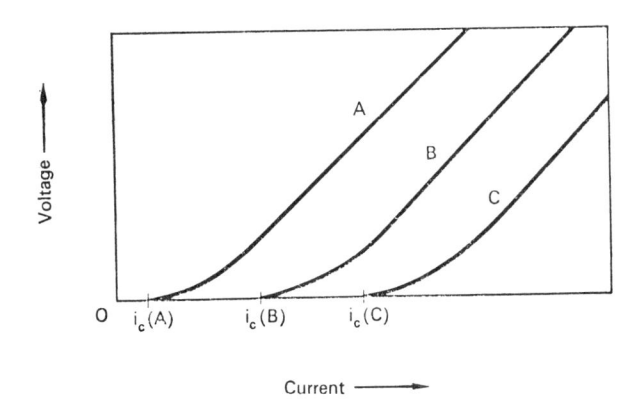

FIG. 13.3. Voltage–current characteristics of three wires of the same type-II superconductor in the mixed state in the same transverse applied magnetic field. Curves A, B, and C refer to specimens which are progressively less perfect.

voltage–current characteristics, measured at the same strength of applied magnetic field, of three wires of equal diameter of the same type-II superconductor but of different degrees of perfection. The critical current is different for each wire, the purer or more perfect wires having lower critical currents; but the slope of the characteristic is the same for all three specimens. We see, therefore, that, though the value of the critical current of a specimen depends on the perfection of the material, the rate at which voltage appears when the critical current is exceeded is an innate characteristic of the particular material and does not depend on how perfect it is.

The value of the slope dV/di which the characteristic approaches at currents well above i_c is known, for reasons which will become apparent later, as the *flow resistance* R' of the specimen. The *flow resistivity* ρ' of

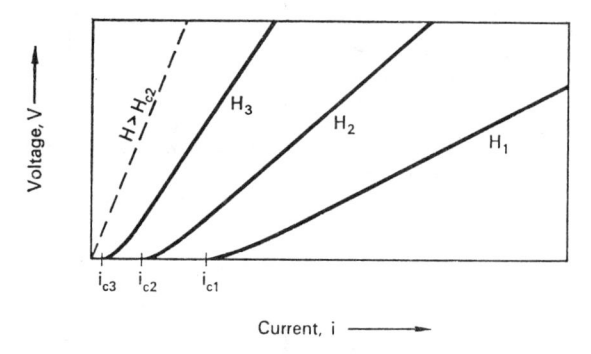

FIG. 13.4. Effect of applied magnetic field strength on V–i characteristic of a type-II superconductor in mixed state in a transverse magnetic field ($H_{c1} < H_1 < H_2 < H_3 < H_{c2}$).

the material from which the specimen is made may be defined by $R' = \rho' l/\mathscr{A}$, where l is the length of the specimen and $\mathscr{A}$ its cross-sectional area. It is found that, for a given strength of applied magnetic field, the flow resistivity is proportional to the normal resistivity of the metal. Furthermore, the flow resistivity increases with increasing strength of applied magnetic field (Fig. 13.4), approaching the normal resistivity as the applied field strength approaches H_{c2}.

13.3. Flux Flow

13.3.1. Lorentz force and critical current

We have seen that a type-II superconductor in the mixed state is able to carry some resistanceless current and that the critical current cannot be determined by any modification of Silsbee's rule. Furthermore, the manner in which voltage appears when the critical current is exceeded is quite different from the case of a type-I superconductor. We now ask what determines the magnitude of the critical current of a type-II superconductor in the mixed state, and what is the source of the voltage which appears at currents greater than the critical current.

The current-carrying properties of type-II superconductors can be qualitatively explained if it is supposed that when a current is passed along a type-II superconductor, which has been driven into the mixed state by an applied magnetic field, the current flows not just at the surface, as in a type-I superconductor, but *throughout the whole body of the metal*.

Consider a length of type-II superconductor in an applied transverse magnetic field of strength greater than H_{c1} (Fig. 13.5). If a current is

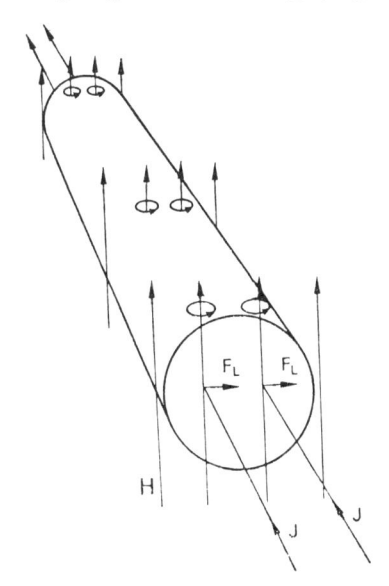

FIG. 13.5. Type-II superconductor carrying current through the mixed state. For stationary cores the Lorentz force F_L is perpendicular both to the axes of the cores and to the current density J.

passed through this specimen, there will be at every point a certain transport current density J. (The "transport current" is the current flowing along a specimen. We use the term to distinguish this motion of the electrons from the circulating vortex currents around the cores.)

However, because the metal is in the mixed state, it is threaded by the magnetic flux associated with the normal cores. There will therefore be an electromagnetic force (Lorentz force) between this flux and the current. When speaking of a Lorentz force between a moving charge and a magnetic field one is in fact referring to the mutual force which exists between the moving charge and the source of the magnetic field. In this case, therefore, the force acts between the electrons carrying the transport current and the vortices generating the flux in the cores. Hence there will be on each vortex a Lorentz force F_L acting at right angles to both the direction of the transport current and to the direction of the flux.

Suppose the specimen is of length l, cross-sectional area $\mathscr{A}$ and carries a current i in an applied magnetic field which is at an angle θ to the direction of the current. If the flux density of the field is B the Lorentz force on the specimen is $liB \sin \theta$. However, since each vortex encloses an amount of flux Φ_0, the mean flux density is $B = n\Phi_0$, where n is the number of vortices per unit area perpendicular to B, and the Lorentz force is therefore $lin\,\Phi_0 \sin \theta$. The total length of all the vortices threading the specimen is $nl\mathscr{A}$, so the mean force per unit length of vortex is $(i/\mathscr{A})\,\Phi_0 \sin \theta$. Though the current density varies between the cores, the average current density $\mathcal{J}$ equals $i/\mathscr{A}$, so the Lorentz force on unit length of each vortex can be seen to be

$$F_L = \mathcal{J}\Phi_0 \sin \theta. \tag{13.1}$$

In the special case where the applied magnetic field is perpendicular to the direction of the current, $\theta = 90°$ and the force is just

$$F_L = \mathcal{J}\Phi_0. \tag{13.2}$$

In the previous chapter we have seen that the cores tend to be pinned at imperfections in the material. Consequently, if the Lorentz force is not too great, the cores remain stationary and do not move under its action. (The electrons which carry the transport current cannot move sideways in the opposite direction, because there can be no component of current across the specimen.) Not every individual core is directly pinned to the material, but the interaction between the vortex currents is sufficient to give the lattice of cores a certain rigidity, so that if only a few cores are pinned the whole pattern is immobilized. What matters, therefore, is the average pinning force per core. Let the average pinning force per unit length of core be F_p. So long as the transport current density $\mathcal{J}$ produces a Lorentz force per unit length of core which is less than F_p, the core lattice will not move, i.e. there will be a stable situation if

$$\mathcal{J}\Phi_0 < F_p.$$

If, however, the transport current is increased, so that the Lorentz force exceeds F_p, the core lattice is no longer prevented from moving through the specimen. If the cores are set in motion,† and if there is some viscous force opposing their motion through the metal, work must be done in maintaining this motion. This work can only be supplied by the transport current, and so energy must be expended in driving this current through the material. In other words, if the current sets the cores into motion, and if their motion is impeded, *there will be a voltage drop along the material*. This motion of the cores (and the fluxons they contain) through the material is known as "flux flow" and is the source of the flow resistivity observed at currents greater than the critical current.

This motion of the cores when the current exceeds the critical current has been observed directly by the effect it has on neutron diffraction from the mixed state‡ (see p. 188).

The mechanisms producing the viscous force that opposed the motion of the cores through the metal are complicated and we shall not discuss them here. One contribution to this viscous drag arises because the cores contain magnetic flux and, as each core moves, this flux moves with it through the metal. This flux motion induces an e.m.f. which drives a current across the core, the current returning via the surrounding superconducting material. These currents may be thought of as eddy currents set up by the flux motion. Since the core is normal, work is dissipated in driving the current across it, and this is one reason why energy must be provided to keep the cores in motion.

It should be stressed that the situation with regard to the voltage is different in type-I and type-II superconductors. In a type-I superconductor, if the critical current is exceeded, the voltage is due to the transport current flowing through normal regions which span the whole specimen. When flux flow is occurring in a type-II superconductor, the material is still in the mixed state and there are still continuous superconducting paths threading the whole specimen.

The critical current will be that which creates just enough Lorentz

---

† When the cores are moving the forces acting on them are different. The velocity and direction of the core motion will be discussed in § 13.3.2.

‡ Schelten, Ullmaier and Lippmann, *Phys. Rev.* **B12**, 1772 (1975).

force to detach the cores from the pinning centres, i.e. the critical current density $\mathcal{J}_c$ will be given by

$$\mathcal{J}_c \Phi_0 = F_P.$$

We can now see why the more imperfect specimens have higher critical currents: if there are many imperfections, a greater fraction of cores will be pinned to the material and the mean pinning force per core will be greater.

In the previous chapter we saw that the presence of pinning centres gives rise to irreversible magnetization of imperfect type-II superconductors. If the above explanation of the critical current is correct, the critical current in the mixed state should be greater in those materials which have more irreversible magnetization curves. This indeed is found to be the case. Figure 13.6 shows magnetization and critical current curves of a wire of type-II superconductor (an alloy of tantalum and

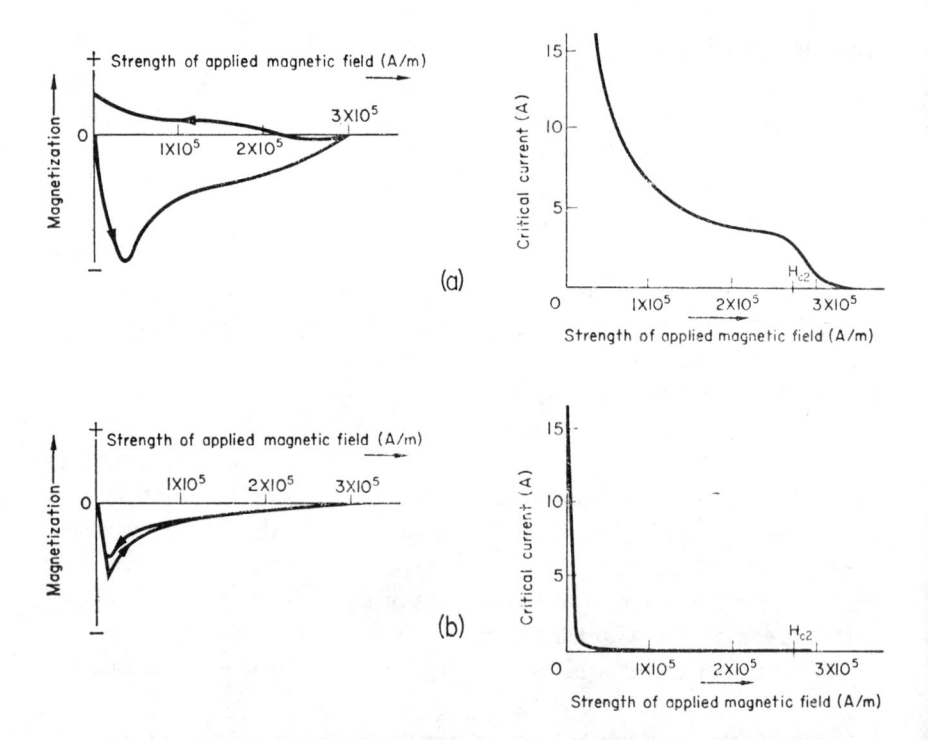

FIG. 13.6. Magnetization and critical current of imperfect (a) and nearly perfect (b) type-II superconductor (tantalum-niobium alloy measured at 4·2°K; Heaton and Rose-Innes).

niobium) before and after it had been purified.† The top two curves show the magnetization and critical current of the wire after it had been drawn from an ingot and so contained many imperfections due to the drawing process. The magnetization curve is very irreversible and there is a plateau of high critical current extending to H_{c2}. The lower curves show the properties of the *same* piece of wire after it had been carefully purified and the imperfections eliminated by heating for several days in a very good vacuum. It can be seen that the magnetization has become almost perfectly reversible and, though the mixed state still persists up to H_{c2}, the high current-carrying capacity has been lost.

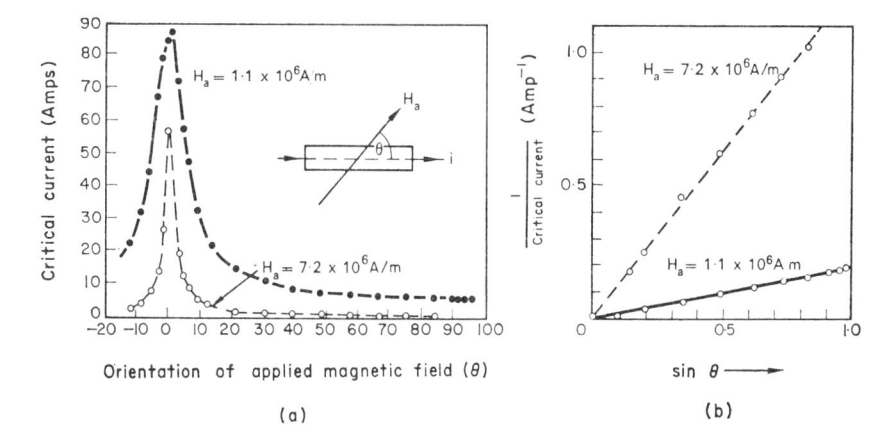

FIG. 13.7. Variation of critical current of Nb₃Sn strip in mixed state with orientation of applied magnetic field. (After G. D. Cody and G. W. Cullen, RCA Laboratories.)

If the critical current is that current which produces a Lorentz force strong enough to move the vortices off any pinning centres, we should expect it to depend markedly on the angle between the magnetic flux and the current. According to (13.1), if θ is the angle the applied magnetic field makes with the current, the critical current should be inversely proportional to $\sin \theta$. Figure 13.7 shows the results of an experiment to test this relationship. It can be seen that, except when the applied field is nearly parallel to the current, the inverse of the critical current does vary

---

† It may be noticed that many of the experiments used to illustrate the properties of type-II superconductors have been performed on tantalum–niobium or niobium–molybdenum alloys. This is because these alloys are two of the few type-II superconductors which can be prepared in a really pure state so that they show reversible magnetization. Though there are many type-II superconducting alloys, most of these cannot, for metallurgical reasons, be produced in very perfect form.

as $\sin\theta$ in accordance with the Lorentz force model. If the inverse of the critical current were to vary as $\sin\theta$ for *all* values of θ, the critical current would tend to infinity as the direction of the applied magnetic field is rotated to lie parallel to the current. Clearly the critical current cannot be infinite, and when the field is parallel to the wire the critical current has a certain maximum value. The factors which limit the critical current is parallel applied magnetic fields are not yet, however, properly understood.

13.3.2. Flux flow

We have associated the voltage, which appears when the transport current is increased above the critical value, with the work required to drive the cores through the metal. For a given current the voltage is independent of time, from which it may be deduced that the core lattice is not accelerating under the forces which act on it but moves with constant velocity. This implies that the metal behaves as though it were a viscous medium as far as motion of the cores is concerned. We have seen

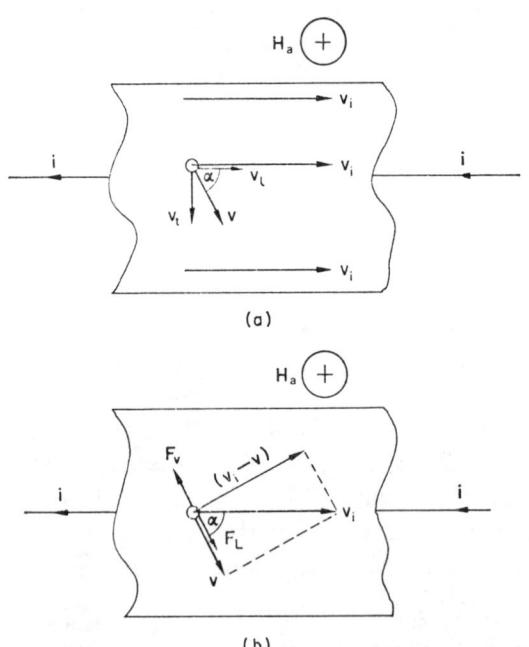

FIG. 13.8. Velocity and force diagrams for vortex motion through mixed state. v_i is the velocity of the electrons carrying the transport current i, and v is the velocity of a vortex through the material. The applied magnetic field is directed into the plane of the page. (Based on Volger, Stass, and van Vijfeiken.)

that the electrical flow resistivity does not depend on the purity of the superconductor, so to each material we can ascribe a "viscosity constant" η such that, if there is a net force F per unit length acting on a core, the core will acquire a velocity $v = F/\eta$.

We now consider what motion the vortices take up when the transport current is raised above the critical value, so that the vortices become detached from the pinning centres. It is important to realize that when the cores are in motion, the magnitude and direction of the forces acting on them are different from when they are held stationary in the material. Consider the special case when the applied magnetic field is perpendicular to the direction of transport current flow. If the cores are held stationary in the metal, the relative velocity between the cores and the electrons carrying the transport current has a certain value, and, as we have seen, there is a Lorentz force tending to detach the vortices from their pinning sites. This force is perpendicular both to the transport current and to the axis of the cores. When, however, the cores have been set in motion, there is a different relative velocity between them and the electrons carrying the current, so the force acting on them is changed. We must remember that the vortices encircling the cores represent a circulatory motion of the superelectrons. This motion is superimposed on the linear motion produced by the transport current, and *in the absence of any other forces* the vortices would be carried along by the transport current with the velocity v_l of its electrons (Fig. 13.8a). If this were to happen there would be no relative motion of the vortices and transport current electrons, and so there would be no Lorentz force on the vortices. However, as we have seen, the appearance of a voltage suggests that there is opposition to the motion of the vortices through the metal, and we now show that, as a result, the vortices acquire a component of velocity at right angles to the transport current.

Because of the viscous drag exerted by the material of the metal on the vortices they will not move as fast as the transport current electrons, but only with some lower velocity v_l. In other words, in the direction of the transport current there will be a relative velocity between the vortices and the electrons carrying the transport current. Consequently there will now be a Lorentz force on the flux threading the vortices, and this gives them a component of velocity v_t sideways across the conductor. The resultant velocity **v** of the vortices is therefore at an angle a to the direction of the transport current (Fig. 13.8a).

We can find the direction of the vortex motion by the following self-consistent argument. If in the steady state the vortices move through the metal with a velocity **v** which is constant, the forces on them must balance. These forces are illustrated in Fig. 13.8b. The metal exerts a viscous drag on vortices moving through it, so there will be a force $\mathbf{F}_v$ acting in a direction opposite to the vortex velocity **v**:

$$\mathbf{F}_v = -\eta \mathbf{v}. \tag{13.3}$$

Now the electrons of the transport current have a velocity $(\mathbf{v}_l - \mathbf{v})$ relative to the core of the vortex and this produces, at right angles to $(\mathbf{v}_l - \mathbf{v})$, the Lorentz force $\mathbf{F}_L$ which drives the vortices in the direction **v** (see Fig. 13.8b). The magnitude of $\mathbf{F}_L$ is given by

$$F_L = |\mathbf{v}_l - \mathbf{v}| n_s e \Phi_0, \tag{13.4}$$

where n_s is the number of superelectrons per unit volume. For the velocity of the vortices to be constant, $\mathbf{F}_L$ must equal $-\mathbf{F}_v$, which from (13.3) and (13.4) gives

$$|\mathbf{v}_l - \mathbf{v}| = \frac{\eta v}{n_s e \, \Phi_0},$$

and the angle α which the vortex motion makes with the transport current is

$$\alpha = \tan^{-1}\left\{\frac{\eta}{n_s e \Phi_0}\right\}.$$

This shows that the greater the viscous drag that the metal exerts on the vortices (i.e. the bigger the value of η) the more nearly will the vortices move at right angles to the transport current. Measurements on type-II superconductors have shown that α is close to 90°. (This result has been obtained from Hall effect measurements. The Hall angle equals $90° - \alpha$.) The fact that α is nearly 90° implies that the viscous drag is large so that when flux flow occurs the vortices move virtually at right angles to the direction of the transport current.

13.3.3. E.M.F. due to core motion

In the previous sections we have seen that a sufficiently strong current i passed through a superconductor in the mixed state sets the cores into motion, and we have supposed that the metal exerts a viscous drag on cores moving through it. The energy required to keep the cores moving can only come from the current and this means that, irrespective of the detailed mechanism of the process, work is required to maintain the current; in other words, there will be a voltage difference V between the ends of the specimen, and the specimen shows resistance. If P is the power required to keep the cores in motion, i.e. the power dissipated in the specimen, the voltage will be simply $V = P/i$.

The above argument as to why a voltage appears at currents greater than the critical current is quite general but gives no insight into the mechanism by which the voltage is generated. In fact, the voltage may be ascribed to an induced e.m.f. generated by the motion of the magnetic flux in the moving cores. Consider, for example, the circuit shown in Fig. 13.9, where M is a piece of type-II superconductor driven into the mixed state by a magnetic field applied perpendicular to the plane of the page. If the cores move across the sample, as the result of a Lorentz force due to a current passed through the sample *or for any other reason*, a voltage V will be recorded on the voltmeter. This voltage is induced by the motion of the magnetic flux contained in the cores. The rate at which flux crosses between the contacts to the voltmeter is $n\Phi_0 v_t d$ where n is the number of cores per unit area, Φ_0 is the magnetic flux within each core (i.e. the fluxon), v_t the transverse velocity of core motion, and d the separation of the contacts. The voltage measured is

$$V = n\Phi_0 v_t d.$$

This "induced voltage" is the same voltage as the "resistive voltage" due to the passage of the current i.

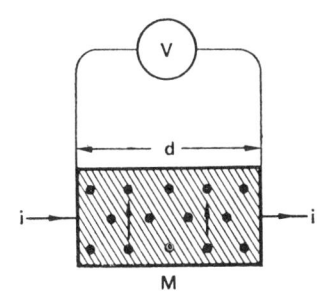

FIG. 13.9. Generation of e.m.f. by flux flow in mixed state.

If it is the motion of the fluxons across a type-II superconductor which produces the e.m.f. in the mixed state, an e.m.f. should appear whenever the cores are set in motion, and the appearance of the e.m.f. should not depend on the cause of the motion. That this is so has been demonstrated by Lowell, Muñoz, and Sousa, who were able to set the fluxons in motion through a specimen of niobium–molybdenum alloy, in which no transport current was flowing, by heating one end of the specimen. If we consider the variation with temperature of the magnetization curve of a type-II superconductor we see that, in a uniform applied magnetic field, there will be a greater density of fluxons in hotter regions than in colder ones. Consequently in a temperature gradient, because of the mutual repulsion of the fluxons, there will be a force driving the fluxons from the hotter to the colder parts. Figure 13.10 shows the arrangement. When a temperature difference was established between the ends of the specimen, a voltage difference V appeared between the two edges. Since no transport current was present, the observed voltage cannot have any "ohmic" source. The voltage changed sign when the direction of the applied magnetic field H_a was reversed. This experiment is the most convincing demonstration that motion of fluxons through the mixed state can generate an e.m.f. The experiment can only be performed on very perfect specimens, otherwise the motion of the fluxon pattern is prevented by the pinning due to imperfections.

We can sum up the current-carrying behaviour of the mixed state as follows: it is supposed that when a current is passed through a type-II superconductor in the mixed state this current flows throughout the metal. This current exerts a Lorentz force on the cores which thread the mixed state. These cores may be anchored to imperfections but if the current exceeds a certain value, the critical current, the cores may be driven across the material. When this "flux flow" occurs a voltage appears perpendicular to the direction of flux motion and heat is generated in the material.

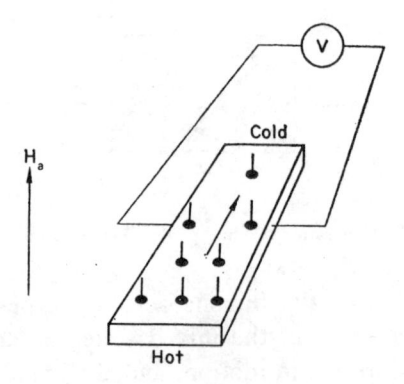

FIG. 13.10. E.M.F. generated by fluxon motion due to a temperature gradient.

It should be made clear that at present (1977) the details of these processes are by no means fully understood. In particular, it is not clear why imperfections pin the core lattice against motion, nor can the form of the variation of critical current with applied magnetic field strength (e.g. Fig. 13.1) be properly explained.

13.4. Surface Superconductivity

It can be seen from Fig. 13.1 that though the critical current of a type-II superconductor falls rapidly when the applied magnetic field strength exceeds H_{c2}, the material is nevertheless able to carry a small, though rapidly decreasing, resistanceless current in fields greater than H_{c2}. This is surprising, because above H_{c2} the metal should be in the normal state and not able to carry any resistanceless current at all. For a number of years, attempts were made to explain this and similar anomalies by supposing that the material was inhomogeneous. It might, for example, be threaded by a network of regions whose critical field was higher than that of the bulk material. It has recently been realized, however, that these "anomalies" are a manifestation of a property which is possessed even by perfectly pure and homogeneous superconductors. This property is that of *surface superconductivity*.

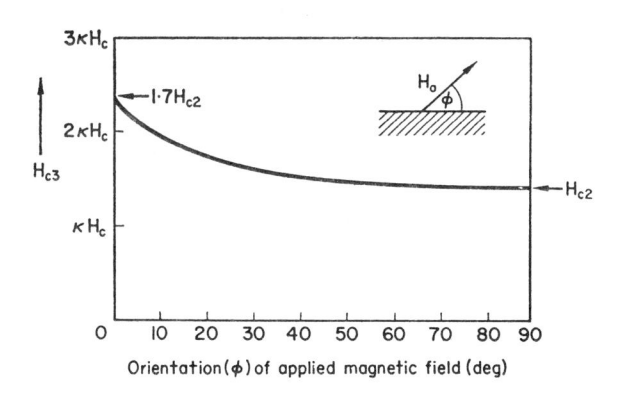

FIG. 13.11. Variation of H_{c3} with angle of applied magnetic field to surface.

In 1963 St. James and De Gennes deduced from theoretical considerations that superconductivity can persist close to the surface of a superconductor in contact with an insulator (including vacuum), even in a magnetic field whose strength is sufficient to drive the bulk material normal. This superconducting surface layer can occur in materials whose

Ginzburg–Landau parameter $\varkappa$ exceeds 0·42. Surface superconductivity is not a special property of type-II superconductors ($\varkappa > 0.71$) but can occur in any superconductor whose $\varkappa$-value exceeds 0·42. However, because its effects are usually observed in type-II superconductors, we have delayed its discussion till this chapter.

The surface superconducting layer† can persist in applied magnetic fields up to a certain maximum strength which we call H_{c3}. The value of H_{c3} depends on the angle the applied magnetic field makes with the surface, and H_{c3} is a maximum when the applied field is parallel to the surface. In this case $H_{c3} = 2 \cdot 4 \varkappa H_c$ (i.e. $1 \cdot 7 H_{c2}$ for a type-II superconductor). If the angle the applied field makes with the surface is increased, the value of H_{c3} decreases (Fig. 13.11) reaching a minimum value of $\sqrt{2\varkappa} H_c$ (i.e. H_{c2} in the case of a type-II superconductor) when the field is perpendicular to the surface.

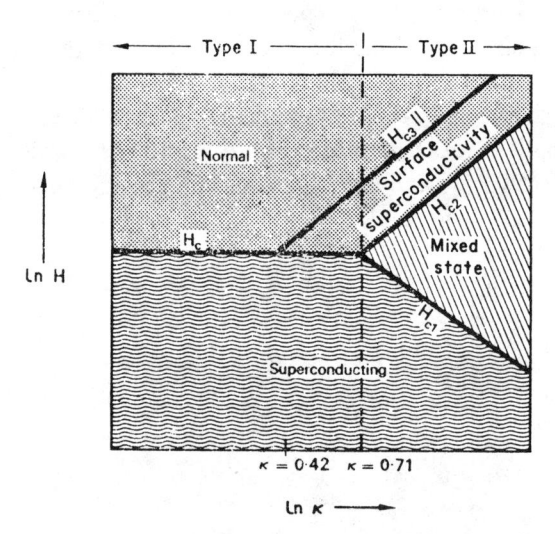

FIG. 13.12. Dependence of characteristics of superconductors on value of Ginzburg–Landau constant $\varkappa$. $H_{c3}||$ is value of limiting field strength for surface superconductivity when field is parallel to surface.

We are now in a position to produce a diagram (Fig. 13.12) which shows how the nature of superconductors depends on the value of their Ginzburg–Landau parameter $\varkappa$. For $\varkappa < 0.42$ the superconductor is type-I and can exist in one of two states, superconducting or normal,

† This surface layer is often referred to as the superconducting surface "sheath". We, however, prefer to speak of the superconducting surface "layer" to emphasize that, as will be soon shown, this layer may appear only on parts of the surface and does not necessarily enclose the specimen.

depending on whether the magnetic field strength is below or above the thermodynamic critical field H_c. However, if $\varkappa$ exceeds 0·42 a thin superconducting layer can exist on the surface in fields up to H_{c3}. When $\varkappa$ is greater than 0·71 a superconductor is type-II and can exist in four possible conditions—superconducting, mixed, normal with surface superconductivity, and completely normal.

It was pointed out in the previous chapter than the $\varkappa$ of pure metals varies slightly with temperature, increasing as the temperature falls. Hence it is possible for a metal to change its type of superconductivity if the temperature is changed. Lead, for example, has a $\varkappa$-value of about 0·37 at 7·2°K, but its $\varkappa$-value increases to 0·58 on cooling to 1·4°K. The value of $\varkappa$ becomes equal to 0·42 at 5·8°K, so at temperatures below this a superconducting surface layer can appear on a lead specimen.

Surface superconductivity occurs only at an interface between a superconductor and a dielectric (including vacuum), and does not occur at an

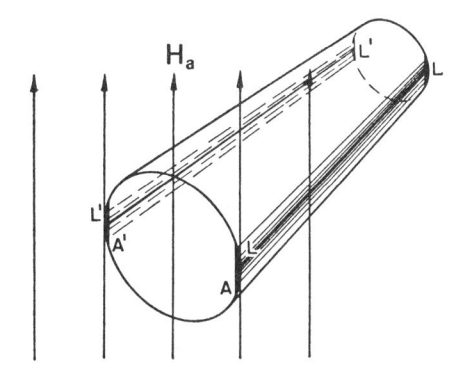

FIG. 13.13. Bands of surface superconductivity along a cylindrical specimen in a transverse magnetic field.

interface between a superconductor and a normal metal. Hence, surface superconductivity can be prevented by coating the surface of a specimen with normal metal. We can, for example, test if an observed behaviour is due to surface superconductivity by seeing whether the behaviour persists after the specimen has been copper-plated.

In general only part of the surface of a specimen is parallel to the applied magnetic field, and consequently, in a magnetic field of strength between H_{c2} and H_{c3}, the superconducting layer will only cover this part of the surface. Figure 13.13 illustrates the important special case of a cylindrical rod or wire in a transverse magnetic field. At an applied field strength equal to H_{c2} the superconducting layer extends right round the surface; but as the field strength is increased, this layer contracts into two bands A and A' running along the sides of the specimen. As the field strength approaches H_{c3} these bands contract to zero along the two lines LL and $L'L'$ where the surface is parallel to the applied field. It is these superconducting bands which account for the small resistanceless current that a type-II superconductor can carry in applied field strengths exceeding H_{c2}.

APPENDIX A

THE SIGNIFICANCE OF THE MAGNETIC FLUX DENSITY B AND THE MAGNETIC FIELD STRENGTH H

IT IS important for a proper appreciation of the magnetic properties of superconductors that the reader should have a sound understanding of the significance of B and H. This is especially so in view of the fact that the MKS system involves a rather different approach to magnetism than the mixed e.s.u.–e.m.u. system.

A.1. Definition of B

The modern approach is to abandon the concept of free magnetic poles and instead to discuss magnetism entirely in terms of the interaction between currents. From this point of view, the magnetic flux density vector **B** is the basic magnetic quantity, in the sense that in electrostatics the fundamental quantity is the electric field vector **E**. If a conductor carrying a current i is placed in a magnetic field a force will act on the conductor due to the presence of the field. The magnetic flux density **B** of the field is *defined* by the relationship:

$$d\mathbf{F} = i d\mathbf{l} \times \mathbf{B}, \tag{A.1}$$

where $d\mathbf{F}$ is the force on a current element $id\mathbf{l}$. This is analogous to the definition of **E** as the force on unit charge, and establishes **B** as the fundamental property.

Magnetic fields are generated by currents, and the flux density resulting from a given geometrical arrangement of currents in free space can be calculated from the Biot–Savart law:

$$d\mathbf{B} = \frac{\mu_0 i \, d\mathbf{l} \times \hat{\mathbf{r}}}{4\pi r^2}, \tag{A.2}$$

where $d\mathbf{B}$ is the contribution to the flux density at a point P due to a current element $id\mathbf{l}$, r is the distance of P from the element $d\mathbf{l}$, $\hat{\mathbf{r}}$ the unit vector in the direction of r, and μ_0 the permeability of free space. This expression for **B** corresponds to the expression

$$d\mathbf{E} = \frac{dq\,\hat{\mathbf{r}}}{4\pi\varepsilon_0\,r^2} \quad (A.3)$$

for the electric field in free space due to a charge dq, ε_0 being the permittivity of free space. It can be shown from the Biot–Savart law that *in free space* **B** satisfies the Ampère circuital law

$$\oint \mathbf{B} \cdot d\mathbf{l} = \mu_0 \mathscr{I}, \quad (A.4)$$

where the line integral $\oint \mathbf{B} \cdot d\mathbf{l}$ is taken round *any* closed path and $\mathscr{I}$ is the net current linking that path.

It can easily be shown by applying the Ampère circuital law to an infinitely long solenoid that the flux density inside it is uniform and given by

$$B = \mu_0 m i, \quad (A.5)$$

where m is the number of turns per unit length and i the current through each of them. Furthermore, the flux density outside the solenoid is zero.

A.2. The Effect of Magnetic Material

All the foregoing equations with the exception of (A.1) apply *only in free space*. To introduce the effect of magnetic material, consider a long cylinder of paramagnetic material inside an infinitely long solenoid, as shown in Fig. A.1. Paramagnetism arises because the material contains

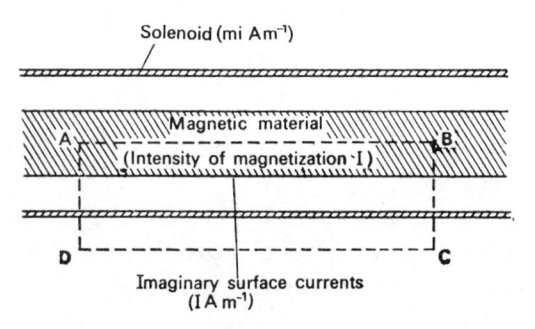

FIG. A.1. Rod of magnetic material in infinite solenoid. $ABCD$ ---- path of integration (A.7).

within it elementary atomic dipoles, and these dipoles tend to be aligned by the field of the solenoid, so that they point predominantly in the direction of the field. Remembering that the atomic dipoles are due not to free

APPENDIX A

magnetic poles, but to small circulating currents which arise either from electron spin or from the orbital motion of electrons within the atoms, it will be seen that these circulating currents are as shown in Fig. A.2a when viewed parallel to the axis of the solenoid. Because of the aligning influence of the field, all these currents circulate in the same sense. The degree of magnetization of the material can be described by specifying its *intensity of magnetization* (usually called simply its "magnetization") **I**, which is a vector pointing in the direction of magnetization and having a magnitude equal to the resultant magnetic dipole moment per unit volume.

The total flux density within the cylinder is now the resultant of the flux density due to the solenoid and that due to the atomic currents.

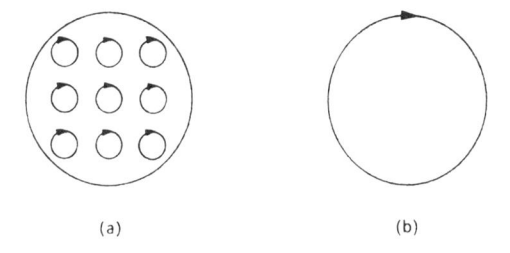

FIG. A.2. Equivalence of aligned current dipoles and surface current (viewed in direction of field). The atomic current loops which generate the magnetic dipoles all circulate in the same direction due to the aligning action of the field, as in (a). The average flux density produced by these current loops within the material is the same as would be produced by imaginary surface currents of density I A m^{-1}, as in (b).

Clearly **B** will not be uniform within the magnetic material but will fluctuate from point to point with the periodicity of the atomic lattice. But there will be an average value of **B**, and it can be shown† that if the magnetic material has a magnetization **I** this average value is exactly the same as would be produced by fictitious currents flowing around the periphery of the cylinder in planes perpendicular to the axis, as in Fig. A.2b, and having a surface density of I A m^{-1}. The magnetization of the paramagnetic material is therefore equivalent to an imaginary solenoid carrying a current of I A m^{-1} and the additional flux density produced by this solenoid is, in accordance with (A.5), given by

$$\mathbf{B}_m = \mu_0 \mathbf{I}.$$

---

† See, for example, A. F. Kip, *Fundamentals of Electricity and Magnetism*, McGraw-Hill, 1962.

The flux density due to this imaginary solenoid simply adds to the flux density produced by the real solenoid, so the magnitude of the total flux density within the material is given by

$$B = \mu_0 mi + \mu_0 I. \qquad (A.6)$$

There are two conventions in the literature about the dimensions of I. Most books on electromagnetic theory give I the dimensions of amperes per metre, as we have done here. Books dealing with the magnetic properties of solids, however, often describe their magnetism in terms of a magnetic *polarization* I', a magnetic *flux density* which the sample adds onto the flux density of the applied field. So I' has the same dimensions as B, and (A.6) becomes

$$B = \mu_0 mi + I'. \qquad (A.6a)$$

There are arguments in favour of each convention, but the definition of I embodied in (A.6) as a magnetic *field strength* contributed by the sample seems to be more in keeping with the parallelism between magnetism and electrostatics, and we have adopted it in this book.

A.3. The Magnetic Field Strength

If we take the line-integral of **B** around the path $ABCD$ in Fig. A.1, where $AB = CD = x$, we find from (A.6), remembering that $B = 0$ outside the solenoid,

$$\oint_{ABCD} \mathbf{B} \cdot d\mathbf{l} = (\mu_0 mi + \mu_0 I)x. \qquad (A.7)$$

We can write this as

$$\oint \mathbf{B} \cdot d\mathbf{l} = \mu_0 (\mathscr{I}_f + \mathscr{I}_m), \qquad (A.8)$$

where $\mathscr{I}_f = xmi$ is the total current through the turns of the solenoid linking $ABCD$, and $\mathscr{I}_m = Ix$ is the total effective surface current which is equivalent to the magnetization I. $\mathscr{I}_f$ is often referred to as the "free" current. Hence (A.4) only remains valid if we identify $\mathscr{I}$ with $\mathscr{I}_f + \mathscr{I}_m$. This is not a useful relationship, however, because although we know $\mathscr{I}_f$ we do not in general know $\mathscr{I}_m$. It is therefore convenient to introduce a new vector, called the *magnetic field strength* **H**, which is defined by

$$\mathbf{B} = \mu_0 \mathbf{H} + \mu_0 \mathbf{I} \qquad (A.9)$$

so that (A.6) gives $\qquad H = mi \qquad (A.10)$

and (A.7) becomes $\oint \mathbf{H} \cdot d\mathbf{l} = \chi m i = \mathscr{I}_f.$ (A.11)

Hence Ampère's circuital law for **H** involves only the *real* or "free" current $\mathscr{I}_f$ and is independent of the presence of the magnetic material. It is this important result which makes **H** a useful quantity. It is an unfortunate feature of the MKS system that the fundamental distinction between B and H, as exemplified by (A.8) and (A.11), tends to be obscured by the presence of a dimensional factor μ_0 in (A.8) which is absent in (A.11).

For many materials (but not iron) it is found that the magnetization I is proportional to the magnetic field intensity *within the specimen*, so that inside the material $\mathbf{I} = \chi \mathbf{H}$ where χ is the susceptibility. Hence

$$\mathbf{B} = \mu_0(\mathbf{H} + \mathbf{I}) = (1 + \chi)\mu_0\mathbf{H}$$
$$= \mu_r\mu_0\mathbf{H},$$

where $\mu_r = 1 + \chi$ is the relative permeability, which is a pure number. For ferromagnetic or paramagnetic materials χ is positive and $\mu_r > 1$, but for diamagnetic materials χ is negative and $\mu_r < 1$.

To sum up, for the case of a long thin rod, the flux density **B** within the rod includes a contribution from the magnetization of the material, while the magnetic field strength **H** does not. In this case the magnetic field strength inside the rod is the same as if the rod were not there.

A.4. The Case of a Superconductor

The foregoing discussion applies also to the case of a type-I superconductor (or a type-II superconductor below H_{c1}), but with an important distinction. The flux density within the superconductor is zero if we ignore penetration effects, and this perfect diamagnetism is brought about by *real* currents which circulate around the periphery of the superconductor; they are in fact the screening currents discussed in Chapter 2.

When it comes to the concept of B and H in a superconductor, there are two ways in which we can proceed. We can focus attention on the screening currents and regard them as real currents not different in nature from the current in the windings of the solenoid. On this view, which regards the bulk of the superconductor as non-magnetic, it is appropriate to rewrite (A.6) in the form

$$B = \mu_0(mi + j_s),$$ (A.12)

where j_s is the surface density of the screening currents per unit length parallel to the axis. The vanishing of B within a superconductor is due to the fact that the terms in brackets in (A.12) are equal and opposite. In this case we have for any closed path

$$\oint \mathbf{B} \cdot d\mathbf{l} = \mu_0(\mathscr{I}_f + \mathscr{I}_s), \qquad (A.13)$$

where $\mathscr{I}_f$ is the total current through the solenoid turns linking the path, and $\mathscr{I}_s$ the total diamagnetic screening current linking the same path. However, although the screening currents are certainly real, we cannot measure them with an ammeter, and we shall continue to define the magnetic field strength $\mathbf{H}$ so that $\oint \mathbf{H} \cdot d\mathbf{l}$ is always equal to the "free" or *measurable* current $\mathscr{I}_f$. This means that we retain (A.10), which states that

$$H = mi, \qquad (A.10)$$

just as in the case of a paramagnetic material, and *the screening currents affect B but not H inside the material.*

Alternatively, it is possible to invert the argument summarized in § A.2 and regard the real currents flowing on the surface of the superconductor as equivalent to fictitious dipoles uniformly distributed throughout the body of the superconductor. (Since the superconductor is diamagnetic, these imaginary dipoles would point the opposite way to the real dipoles in a ferromagnetic or paramagnetic specimen.) From this point of view, we can talk about the intensity of magnetization I of a superconductor and regard this either as the magnetic moment per unit volume of the equivalent dipoles or the surface density of the screening currents in amps per metre. A practical definition of I is that it is the total magnetic moment of the specimen arising from the surface currents divided by the volume of the specimen. I is equal to and has the same dimensions as the quantity j_s occurring in (A.12). In terms of I, we may write, in accordance with (A.6),

$$B = \mu_0(mi + I), \qquad (A.6)$$

where the term $\mu_0 mi$ represents the flux density due to the solenoid and $\mu_0 I$ is the flux density resulting from the fictitious equivalent dipoles. As before, we write

$$\mathbf{B} = \mu_0 \mathbf{H} + \mu_0 \mathbf{I}, \qquad (A.9)$$

and the vanishing of B implies that within the superconductor the field strength H must be equal to $-I$, so that we have $\chi = -1$ or $\mu_r = 0$, i.e. the bulk of the superconductor is perfectly diamagnetic.

Whichever way we choose to look at it, we have

$$\oint \mathbf{H} \cdot d\mathbf{l} = \mathscr{I}_f \neq \frac{1}{\mu_0} \oint \mathbf{B} \cdot d\mathbf{l}, \tag{A.11a}$$

and the value of H inside a long thin superconductor is the same as if the superconductor were not there. (H is in fact the quantity we have called the applied field H_a). Whichever view we adopt, we say that deep within a superconductor $B = 0$, but in the presence of an applied field H does not vanish.†

For most purposes the second approach is more convenient, since it allows us to apply to a superconductor concepts such as energy of magnetization and demagnetizing field which were first developed for the case of ordinary magnetic materials. As an example, the treatment of the intermediate state given in Chapter 6 would be very difficult if we had to take the screening currents into account explicitly. There are, however, cases where we are particularly interested in the spatial distribution of flux density (and of screening currents) near to the surface of a superconductor. An example is the development of the London equations in Chapter 3. In this case, it is advantageous to adopt the first approach and recognize the existence of the real surface currents which circulate around the specimen whose bulk consists of non-magnetic material. The distinction between B and H is now revealed by (A.11a) and (A.13), which in point form become

$$\operatorname{curl} \mathbf{H} = \mathbf{J}_f \quad \text{and} \quad \operatorname{curl} \mathbf{B} = \mu_0 (\mathbf{J}_f + \mathbf{J}_s).$$

Within the superconductor, $\mathbf{J}_f = 0$, so that $\operatorname{curl} \mathbf{H} = 0$ and $\operatorname{curl} \mathbf{B} = \mu_0 \mathbf{J}_s$.

Both approaches are encountered in the literature, and the reader should familiarize himself with each of them.

A.5. Demagnetizing Effects

So far we have limited our discussion to the case of a long thin specimen in which end effects are unimportant. We now show that the

---

† This standpoint is not invariably taken in the literature. It is quite common to find authors who write $B = \mu_0 H$ everywhere within the superconductor. In this case the vanishing of B implies the vanishing of H also. But there is really no point in introducing B and H if they are everywhere simply related by a universal constant of proportionality. This way of looking at it causes difficulties in the explanation of demagnetizing effects, and also means that the usual boundary condition involving the continuity of the tangential component of H at the boundary between two media, which we have made use of in Chapter 6, is no longer valid at the interface between the superconducting and normal phases.

relationship $B = \mu_0(H + I)$ has to be interpreted somewhat differently if the specimen is not long in comparison with its width. To be specific, consider a sphere of paramagnetic material which is placed in a uniform applied field, say within a long solenoid. It is possible to show that the magnetization **I**, defined as the magnetic moment per unit volume, is uniform within the sphere and is parallel to the axis of the solenoid. It can also be shown[†] that the flux density due to the magnetization is the same as would be produced by surface currents having a constant surface density *I per unit length measured parallel to the axis*. The flux density produced by such a distribution of surface currents around the surface of the sphere is equal to $\frac{2}{3}\mu_0 I$, so within the sphere the flux density is uniform and given by

$$B_i = \mu_0 mi + \tfrac{2}{3}\mu_0 I. \tag{A.13}$$

Comparing this with (A.6) and (A.9), which state that for a long thin specimen

$$B = \mu_0 mi + \mu_0 I = \mu_0 H + \mu_0 I,$$

we might be tempted to define the magnetic field strength by

$$B = \mu_0 H + a\mu_0 I,$$

where a is a numerical factor depending on the geometry of the specimen, equal to unity for a long thin specimen and to 2/3 for a sphere. If we were to do this, the field strength inside the body would be equal to mi both for a long thin specimen and for a sphere. However, unlike the case of a long thin rod, in the case of the sphere the surface currents alter the flux density (and therefore the field strength, since $B = \mu_0 H$) *outside* the sphere, and an argument identical with that given on page 65 (Chapter 6), shows that we should no longer have $\oint \mathbf{H} \cdot d\mathbf{1} = \mathscr{I}_f$. It can be shown that if we wish to retain $\oint \mathbf{H} \cdot d\mathbf{1} = \mathscr{I}_f$ for any closed curve, where $\mathscr{I}_f$ is the total current in those turns of the solenoid which link the curve, then we must also retain (A.9) as the definition of the magnetic field strength $\mathbf{H}_i$ inside the sphere, i.e.

$$\mathbf{B}_i = \mu_0 \mathbf{H}_i + \mu_0 \mathbf{I}. \tag{A.9'}$$

But, according to (A.13),

$$B_i = \mu_0 mi + \tfrac{2}{3}\mu_0 I. \tag{A.13}$$

[†] See, for example, A. F. Kip, *Fundamentals of Electricity and Magnetism*, McGraw-Hill, 2nd ed., 1969, p. 370.

APPENDIX A

These equations can both be satisfied only if

$$H_i = mi - \tfrac{1}{3}I,$$

i.e. the magnetic field strength inside the sphere is different from the value mi which it would have if the sphere were absent and which we have called the applied field H_a. Hence for a sphere

$$H_i = H_a - \tfrac{1}{3}I,$$

and for a body of arbitrary shape

$$\mathbf{H}_i = \mathbf{H}_a - n\mathbf{I}, \tag{A.14}$$

where n is the "demagnetizing factor", equal to 1/3 for a sphere. For a superconductor, if we adopt the approach which ignores the screening currents but regards the bulk of the superconductor as perfectly diamagnetic, then

$$B_i = \mu_0 H_i + \mu_0 I = 0,$$

so that

$$I = -H_i$$

and from (A.14) $H_i = H_a + nH_i$, i.e. the field strength inside the specimen is *increased* to $H_i = H_a/(1-n)$. For the general case we may write

$$\mathbf{B}_i = \mu_0 \; (\;\; \mathbf{H}_a \;\;\; - \;\;\; n\mathbf{I} \;\;\; + \;\;\; \mathbf{I} \;\;)$$

$\qquad\qquad\qquad\;\;\downarrow\qquad\quad\;\;\downarrow\qquad\qquad\downarrow$

applied   demagnetizing   magnetization
field   field

internal field, $\mathbf{H}_i$

APPENDIX B

FREE ENERGY OF A MAGNETIC BODY

SUPPOSE that a body of volume V is uniformly magnetized by an applied field H_a, so that its magnetic moment is M. The field is now increased to $H_a + dH_a$ and produces a change of magnetic moment dM. The total energy which must be supplied (e.g. by a battery driving a coil which generates the magnetic field) to make these increments in field and magnetic moment at constant temperature is[†]

$$dW_{\text{tot}} = V \cdot d(\tfrac{1}{2}\mu_0 H_a^2) + \mu_0 H_a dM.$$

The first term on the right-hand side represents the work done in increasing the strength of the applied field over the volume occupied by the body; this work must be done to increase the field whether the body is present or not. The second term, $\mu_0 H_a dM$, is the energy which must be supplied to increase the magnetic moment of the body. Let us call this dW_M,

$$dW_M = \mu_0 H_a dM.$$

Comparison of this with the work done by an external pressure p in producing an infinitesimal change in volume dV of a body,

$$dW_V = -pdV,$$

shows that the expression for the work to magnetize the body has a similar form, if we suppose that $\mu_0 H_a$ corresponds to p and M to $-V$.[‡]

Now the Gibbs free energy for a body in the absence of a magnetic field is

$$G = U - TS + pV,$$

[†] See, for example, A. B. Pippard, *Classical Thermodynamics*, Cambridge University Press, 1957, p. 26.

[‡] The signs are different because energy must be given *to* a body to increase its magnetization, whereas work is done *by* a body when it increases in volume against an external pressure.

where U and S are its internal energy and entropy. As we might expect from the above considerations, the effect of magnetization is to add a term $-\mu_0 H_a M$ analogous to the term $+pV$:

$$G = U - TS + pV - \mu_0 H_a M.$$

Small changes in the conditions will produce a change in G given by

$$dG = dU - TdS - SdT + pdV + Vdp - \mu_0 H_a dM - \mu_0 M dH_a.$$

If the applied field H_a and the magnetic moment M are changed while the temperature and pressure are kept constant ($dT = dp = 0$), we have

$$dG = dU - TdS + pdV - \mu_0 H_a dM - \mu_0 M dH_a.$$

But for a magnetic body under the same conditions of constant T and p,

$$dU = \underbrace{TdS - pdV + \mu_0 H_a dM}_{\text{work done on body}}.$$

Therefore $dG = -\mu_0 M dH_a$, and the change in free energy of a body when it is magnetized to a magnetic moment M by a field of strength H_a is

$$G(H_a) - G(0) = -\mu_0 \int_0^{H_a} M dH_a.$$

INDEX

Adiabatic magnetization 61
A. C. Josephson effect 150, 165
Alloys xvii, 6, 185
  thermal propagation 87
  type-II superconductivity xvii, 186, 211
  upper critical fields 195

Bardeen–Cooper–Schrieffer theory (BCS theory) 117–139
  criterion for superconductivity 134
  critical current density 138
  critical magnetic field 134
  current-carrying states 135–139
  energy gap 129–133, 142, 151
  fundamental assumptions 125
  ground state 125
  latent heat 133
  law of corresponding states 134
  macroscopic properties 131–135
  second-order transition 133
  transition temperature 131
  two-fluid model 139
  zero resistance 137
Boundary between superconducting and normal regions 68, 80
  boundary conditions for **B** and **H** 69

Circuits, resistanceless 8,
Coherence 77, 153
  BCS theory 128
  of electron pairs 128, 138, 153
  and surface energy 77, 185
Coherence length 77, 114, 128
  and electron mean free path 79
  order of magnitude 78
  and purity 78
  and surface energy 80, 185
  temperature-dependent 80
  temperature-independent 80
Cooper pairs *see* Electron pairs

Cores 186
  motion 209, 212–216
  pinning 199, 208
Correlation between electrons 80, 129
Corresponding states 134
Coulomb repulsion between electrons 117, 120, 133
Coupling energy 167
Criterion for superconductivity, BCS theory 134
Critical current 40, 82–91, 202–204, 206–212
  density *see* Critical current density
  effect of magnetic field 82, 84–85, 202–206, 210–212
  and irreversible magnetization 210
  Josephson junction 150, 161
  measurement of 87, 91
  quantum interferometer 173, 175, 178, 179
  Silsbee's hypothesis 83
  thin specimens 106–111
  type-II superconductors 202–204, 206–212, 217
  weak link 169, 178
  wires 83
Critical current density 40, 82, 107, 111, 138
  BCS theory 138
  and critical magnetic field 83, 107
Critical magnetic field 40–53
  BCS theory 134
  of cylinder 97
  effect of penetration 92
  elements 6
  H_{c3} 218
  lower critical field H_{c1} 191, 196
  parallel-sided plate 93
  paramagnetic limit 195
  temperature dependence 43
  thermodynamic value 43, 193, 198
  thin film 97–99

Critical magnetic field—*continued*
  thin wire 97
  type-II superconductors 191–196
  upper critical field H_{c2} 192, 194, 196
Critical temperature *see* Transition temperature
Cryotron 45
Crystal structure and superconductivity 113
Current-carrying states, BCS theory 135
Currents in superconductors 11, 22, 26, 36, 202
  analogy with electrostatic charge 23
  critical *see* Critical current *and* Critical current density
  resistanceless networks 11
  transport 23, 82, 192

Defects *see* Imperfections
Demagnetization 66, 227–229
Demagnetizing factor 64, 66, 105, 229
Demagnetizing field 66, 229
Density of states at Fermi surface 127, 134
Diamagnetism, perfect 17, 21, 160
Dislocations 199

Edge effects in magnetic transitions 104
Effective momentum in magnetic field 102
Electric field in superconductor 13, 14, 32, 35
Electrical resistance
  maximum value in superconductor 8
  metals and alloys 3, 112
  at optical frequencies 14, 112
  residual 4, 5
  restoration by current 89
  zero resistance 3, 8, 31, 83, 137, 202
Electrical resistivity
  a.c. 12
  residual 4
  of a superconductor 10
Electron–electron interaction 117
  coulomb repulsion 117, 120, 133
  by electron–phonon interaction 118–120
Electron-lattice interaction 118–120
  conservation of energy and momentum 119–120
  by phonon emission 118
Electron pairs (Cooper pairs) 120, 158
  Bose–Einstein statistics 126
  energy to break up 130
  formation of ground state 121, 123
  long-range coherence 138, 153
  momentum pairing condition 123
  motion of centre of mass 136
  and Pauli principle 126, 127
  phase coherence 128, 139, 154
  spatial extent 129
  total momentum 135
  wave function 124, 136
Electron-pair waves 124, 136, 153, 154, 157
  coherence 153, 154
  diffraction 178
  effect of magnetic field 155
  frequency 154
  interference 170, 178, 179
  momentum 154
  phase 154, 155, 157, 160
  wavelength 153, 154, 155
Electron–phonon interaction *see* Electron-lattice interaction
Electrons
  Fermi "sea" 120
  normal 12, 138
  ordering of 60
  superelectrons 12, 138
  in vacuum 14
Elements
  critical fields 6
  superconducting xvii, 6
  transition temperatures 6
Energy conservation in phonon emission or absorption 119
Energy gap 61, 115, 129–132, 140
  absorption of electromagnetic radiation 115
  BCS theory 130–132, 142, 151
  definition of Δ 127
  measurement of 132, 145, 150
  and specific heat 61, 116
  temperature dependence 132
Energy level diagram 142
  semiconductor representation 147
Entropy 54
  type-II superconductors 201

Fermi "sea" 120
Ferromagnetics 5
Flow resistance 204
Flux *see* Magnetic flux
Flux flow 206, 212
  due to temperature gradient 216
  e.m.f. 214
Flux lines *see* Fluxons
Fluxoid 156–160
Fluxon lattice 187

INDEX

Fluxons 158, 171, 189
   arrangement in mixed state *Frontispiece*, 187, 197
Free energy *see* Gibbs free energy

Gapless superconductors 116, 137, 139
Gibbs free energy 41–43, 70, 230–231
   and critical magnetic field 41–44, 93
   effect of magnetic field 41, 191, 231
   in intermediate state 70
Ginzburg–Landau constant, $\varkappa$ 190, 198, 218
   dependence on resistivity 191
   H_{c3} 218
   surface superconductivity 217
   temperature variation 191, 219
   and type of superconductor 190, 218
Ginzburg–Landau theory 101
   critical magnetic field of thin films 103
   effect of magnetic field 102
   second-order transition in thin film 103
Ground state, BCS theory 125

Hall effect in mixed state 214
High frequencies 14, 113
Hollow superconductor 24, 159
Hysteresis, 48, 199, 210

Ideal superconductor 47
Imperfections 48, 199, 211
Inductance of a superconductor 13–14
Intensity of magnetization *see* Magnetization
Interferometer 170
Intermediate state 66–81
   domain structure 72
   experimental observation of 72
   induced by current 89
   laminar structure 76
   magnetic properties 69
   size of domains 74
   thin films 105
Internal magnetic field 64, 69, 228–229
Isotope effect 115, 133

Josephson tunnelling 149, 160
   a.c. effect 150, 165
   coupling energy 167
   junction 149, 160, 162
   pendulum analogue 161

Kappa, $\varkappa$ *see* Ginzburg-Landau constant, $\varkappa$

Laminar structure of intermediate state 68–70, 76
Latent heat 58, 200
   BCS theory 133
Law of corresponding states 134
London equations 34, 37
London theory 33–39
   limitations of 98
   penetration depth 35
Long-range order 77, 114
   and coherence length 77, 114, 128–129
Lorentz force on fluxons 208

Magnetic field
   "applied" 16
   critical *see* Critical magnetic field
   effect on critical current 84, 203, 206
   effect on electron-pair wave 155
   effect on free energy 41–43, 191, 231
   effect on Josephson junction 178
   internal 66, 69, 228–229
   quantum interferometer 170–180
Magnetic field strength, definition 224
Magnetic flux
   arrangement in mixed state *Frontispiece*, 186–189
   distribution inside type-I superconductor 36, 38
   distribution in parallel-sided plate, 39, 93
   fluxoid, 156–160
   penetration 26
   in perfect conductor 16
   quantum of *see* Fluxons
   in resistanceless circuit 8–11
   trapped 48, 199
Magnetic flux density 46
   definition of 221
   measurement of 49
   zero in superconductor 19
Magnetic moment
   effect of penetration 95
   parallel-sided plate 95
Magnetic shielding by superconductors 11
Magnetization 46
   adiabatic 59
   definition of 223, 224, 226
   hysteresis 48, 198, 211
   irreversible 48, 199, 210
   measurement of 50–53
   parallel-sided plate 95
   "perfect" conductor 16
   superconductor 19

Magnetization—*continued*
  type-II superconductor 197–199, 210
Matrix element of scattering interaction 123, 127
Maxwell's equations 32
Meissner effect 19, 34, 115
  see also Perfect diamagnetism
Mercury xvi
Microscopic theory of superconductivity 112–139
Mixed state 184, 186–190, 191–193, 197–198
  critical current 202–212
  current-carrying properties 192–220
  e.m.f. due to core motion 214
  flux flow 206–217
  fluxon pattern *Frontispiece*, 188, 197
  free energy 189, 192
  Hall effect 214
  lower critical field H_{c1} 191, 196
  upper critical field H_{c2} 192, 194–196
Modulus of elasticity, change at transition 62
Momentum
  conservation in phonon emission or absorption 118
  of electron pair 135–137, 153

Network, resistanceless 11
Neutron diffraction from mixed state 188, 209
Niobium
  transition temperature 6, 7
  type-II superconductivity 6, 191
Non-ideal superconductor 47, 199, 200
Normal electrons 12, 138

Order, of electrons 60

Paramagnetism 195
Peltier coefficient 63
Pendulum analogue 162
Penetration depth 26–30, 98, 160
  dependence on magnetic field 100
  dependence on specimen size 100
  effect on critical magnetic field 92
  effect on magnetization 93
  London theory 35, 37, 99
  and surface energy 80, 185
  temperature variation 28, 36
"Perfect" conductor 16, 20
  magnetic behaviour 16

Perfect diamagnetism 17, 21, 26, 28, 34, 160
  see also Meissner effect
Permeability of superconductor 21
Persistent current 8, 10
Phase *see* Electron-pair wave
Phase coherence of Cooper pairs 128, 138, 153
Phase diagram 43, 194
Phase transition
  first order 59
  second order 57, 200
Phonons
  average frequency 123, 127
  electron-lattice interaction 117
  virtual emission 119, 123
Polymer, superconducting 6

Quantized magnetic flux *see* Fluxon
Quantum interference 153, 170, 180
Quantum interferometer 170, 180
  diffraction effects 179
  interference pattern 173, 178, 180
Quasiparticles 130, 131
  energy of 143

Resistance *see* Electrical resistance
Resistance transitions 4, 8, 83–91, 205
  role of surface energy 76
  temperature range 8
  in transverse magnetic field 76
  zero 83, 138, 202
Resistanceless conductor *see* Perfect conductor
Resistivity *see* Electrical resistivity
Rutger's formula 58

Screening currents 17, 21, 25
Second-order transition 57, 200
  BCS theory 133
  of thin film in magnetic field 103, 152
Silsbee's hypothesis 83–86
  thin specimens 106, 110
  type-II superconductor 203, 204
Small specimens 92
Solenoid, superconducting 10, 195, 202
Sommerfeld specific heat constant 60, 134
Specific heat 57–61, 114
  electronic 59, 114
  and energy gap 61, 116
  lattice 59, 114
  Rutgers' formula 58

Specific heat —*continued*
  Sommerfeld constant 60, 134
  type-II superconductors 200
SQUID *see* Quantum interferometer
Superconducting state
  summary of properties 112
Superelectrons 12, 138
Surface energy 74, 183
  coherence range 77, 185
  and domain size in intermediate state 74
  effect on magnetic field at boundary 76
  measurement of 76, 80
  negative 184, 185–186
Surface superconductivity 217
Susceptibility of superconductor 21
Switch, thermal 2
Symbols xii

Technetium, type-II superconductor 6, 191
Thermal conductivity 62
Thermal expansion, change at transition 62
Thermal propagation 86–88, 111
Thermal switch 62
Thermodynamic variables 41
Thermodynamics 54–63, 230
Thermoelectricity 63, 216
Thin films 28, 93–100, 103–111
  critical magnetic field 93–98, 196
  current distribution 107–109, 110
  intermediate state 106
  thermal propagation 111
  transition in perpendicular magnetic field 105
Thomson coefficient 63
Transition
  first-order 59
  latent heat 58
  reversible 40, 41
  second-order 57, 103, 152, 200
  sharpness of 8, 79
Transition temperature
  alloys and compounds 7
  BCS theory 131
  elements 6
  and energy gap 132
  isotope effect 115, 133
  niobium 7
  range 8
  very low 5
Transport current *see* Currents in superconductors

Tunnelling 140–152, 160, 167
  a.c. Josephson effect 150, 165
  attenuation length 141
  between dissimilar superconductors 149
  between identical superconductors 145
  between normal metals 140
  between normal metal and superconductor 143
  conditions for 141
  experimental details 150
  Josephson 149, 160, 167
  measurement of energy gap 145
  wavefunction 141
Two-fluid model 13, 77, 114
  BCS theory 138
Type-I superconductors xvii, 3–180, 184, 190, 218
Type-II superconductors xvii, 183–220
  alloys 186
  critical current 202–220
  current-carrying properties 207
  e.m.f. due to core motion 209, 214
  flow resistance 206, 209
  Ginzburg–Landau parameter 190
  Hall effect 214
  intrinsic 191
  lower critical field H_{c1} 191, 196
  magnetization 197, 199
  mixed state 183, 186
  specific heat 200
  thermodynamic critical field 193, 198
  thermoelectric effects 63, 216
  upper critical field H_{c2} 192, 194, 196

Vanadium, type-II superconductor 6, 191
Virtual emission of phonon 119, 123
Voltage due to flux flow 214
Volume change at transition 62
Vortices *see* Cores

Wavefunction
  BCS ground state 126
  electron pairs 124, 135
Wavelength of electron-pair wave 153, 155
Waves, electron-pair 153
Weak link 160, 169
  critical current 161, 169, 178
  diffraction 178